The Legal Status of the Caspian Sea

Barbara Janusz-Pawletta

The Legal Status of the Caspian Sea

Current Challenges and Prospects for Future Development

Second Edition

Barbara Janusz-Pawletta
Kazakh-German University
Almaty, Kazakhstan

ISBN 978-3-662-63542-1   ISBN 978-3-662-63540-7   (eBook)
https://doi.org/10.1007/978-3-662-63540-7

© Springer-Verlag GmbH Germany, part of Springer Nature 2015, 2021
This work is subject to copyright. All rights are reserved by the Publisher, whether the whole or part of the material is concerned, specifically the rights of translation, reprinting, reuse of illustrations, recitation, broadcasting, reproduction on microfilms or in any other physical way, and transmission or information storage and retrieval, electronic adaptation, computer software, or by similar or dissimilar methodology now known or hereafter developed.
The use of general descriptive names, registered names, trademarks, service marks, etc. in this publication does not imply, even in the absence of a specific statement, that such names are exempt from the relevant protective laws and regulations and therefore free for general use.
The publisher, the authors, and the editors are safe to assume that the advice and information in this book are believed to be true and accurate at the date of publication. Neither the publisher nor the authors or the editors give a warranty, expressed or implied, with respect to the material contained herein or for any errors or omissions that may have been made. The publisher remains neutral with regard to jurisdictional claims in published maps and institutional affiliations.

This Springer imprint is published by the registered company Springer-Verlag GmbH, DE part of Springer Nature.
The registered company address is: Heidelberger Platz 3, 14197 Berlin, Germany

Preface

The Caspian Sea is a great ecologic, economic, political, and until recently legal chunk of uncertainty worldwide. For almost 30 years, it posed a challenge not only to its five bordering states—Azerbaijan, Iran, Kazakhstan, Russia, and Turkmenistan—but also to the international community interested in its long-term sustainable development in the region and mutually beneficial economic cooperation. This ambitious intention could have been met only by way of interstate cooperation in all dimensions of the development of the Caspian Sea—political, economic, ecological, and also legal. In recent decades, following the collapse of the Soviet Union and the new geopolitical situation in the region, the balance between economic profits from the development of mineral resources on one hand and the protection of the natural environment on the other was not achieved in a sufficient way. Finally, in 2018, the Convention on the Legal Status of the Caspian Sea (further as Caspian Sea Convention) was concluded after years of interstate negotiations. It offers a reason to believe that the long-awaited consensus on the sustainable development of the Caspian Sea has been found.

The presented book represents an attempt to analyze the problem of the international legal status of the Caspian Sea. Until recently, the unclear legal situation of the Caspian Sea, the consequent uncertainty of the coastal states about the issue of territorial demarcation, their uncertainty about the extent of their sovereign rights to the exploitation of natural resources, and the uncertainty of the neighboring states with regard to shipping in the Caspian prevented continuous economic development of the region, destabilized political situation, and resulted in a lack of security in the Caspian region. These issues were not merely of regional but also of global importance. Since 2018, however, when the Caspian Sea Convention was finally adopted, there has been a need for a comprehensive look at the newly adopted interstate legal framework, which is offered in this second edition of the book.

This statement brings me to the sincere wish to express my deepest gratitude to international scholars. Thanks to their profound knowledge and expertise in the Caspian issues, they inspired me and led me through the sometimes difficult way of exploring and assessing the legal status of the Caspian Sea: Dr. Friedemann Müller of the German Institute for International and Security Affairs (SWP) Berlin, Prof. Dr. Philip Kunig of Free University Berlin, Prof. Dr. hab. Leonard Łukaszuk of the University of Warsaw, Prof. William E. Butler of the Pennsylvania State University, Prof. Alexander N. Vylegzhanin of Moscow State Institute of International Relations (MGIMO), and late Prof. Anatoly L. Kolodkin. I also wish to thank my assistant, Angelina Funtikova, for her support in the process of editing the book.

Almaty, Kazakhstan   Barbara Janusz-Pawletta

Contents

1	**Introduction and Course of the Investigation**..............	1
	1.1 Situation in the Caspian Sea After the Collapse of the Soviet Union......................................	1
	1.2 Adoption of the Caspian Sea Convention Defines the Importance of Research............................	2
	1.3 Present State of Research Into the Legal Status and Regime of the Caspian Sea.................................	4
	1.4 Structure of the Book.................................	5
	References..	8
2	**Geography, Politics, and Economy in the Caspian Region**......	11
	References..	16
3	**Legal History and the Present State of Use of the Caspian Resources**...	17
	3.1 Introduction..	17
	3.2 Russian–Persian Treaties Concluded Until the Nineteenth Century...	19
	3.3 Legal Heritage of the Twentieth Century in the Caspian Region..	20
	3.4 State Succession and the Soviet–Iranian Agreements.......	22
	3.5 Legal Interpretation of the Soviet–Iranian Treaties..........	25
	3.5.1 Caspian Sea as a "Sea".......................	26
	3.5.2 Caspian Sea as a "Lake" in Legal Terms..........	28
	3.5.3 The Caspian Sea as "Condominium" in Legal Terms.......................................	30
	3.5.4 North Caspian Agreements.....................	32
	3.6 Legal Confusions in State Practice Regarding the Use of Resources in the 1990s..............................	32
	References..	35

4 Cooperation Levels in Caspian States Practice in the 1990s Until 2018 37
4.1 Challenges for the Caspian Region After the Dissolution of the Soviet Union 37
4.2 Peaceful Settlement in International Law 39
4.3 Five-Party Negotiations on the Convention on the Legal Status of the Caspian Sea from 1990 Till 2018 41
4.4 Step-by-Step Conclusion of Agreements on the Use of Natural Resources 45
4.5 Step-by-Step Multilateral Regulations of the Legal Regimes in the Caspian Sea 47
    4.5.1 Protection of the Marine Environment of the Caspian Sea 48
    4.5.2 Aquatic Biological Resources of the Caspian Sea . . . 51
    4.5.3 Cooperation on the Prevention and Elimination of Emergency Situations in the Caspian Sea 52
    4.5.4 Security Cooperation in the Caspian Sea 52
4.6 Current Legal Framework for the Status of the Caspian Sea as Defined in the Caspian Sea of 2018 54
    4.6.1 General Principles for Interstate Cooperation 55
    4.6.2 Security Cooperation According to the Caspian Sea Convention 57
4.7 Conclusion 57
References 58

5 Interrelations Between Territorial Delimitation and the Regime of the Use of the Caspian Sea 61
5.1 Nonlegal Aspects of Settlement of the Seaward Boundaries in the Caspian Sea 61
5.2 Territorial Delimitation and State Sovereignty 65
5.3 State Practice in the Delimitation of the Caspian Sea Until 2018 67
5.4 Delimitation of the Caspian Sea as Reflected in the Multilateral Negotiations Among the Caspian States Until 2018 71
5.5 Territorial Division of the Caspian Sea According to Caspian Sea Convention 74
    5.5.1 Internal Waters 76
    5.5.2 Territorial Waters in the Caspian Sea Convention . . . 77
    5.5.3 The Fishery Zone and the Common Maritime Space 78
    5.5.4 Methods of Maritime Delimitation of the Caspian Sea 81
5.6 Conclusion 83
References 85

6	**The Regime for the Use of Nonliving Resources in the Caspian Sea**..		87
	6.1	Reserves of Nonliving Resources in the Caspian Sea.......	87
	6.2	International Legal Regulation of Nonliving Resources.....	88
	6.3	Claims on the Rights to Use Nonliving Resources in the Caspian Sea..	89
	6.4	Use of Nonliving Resources in the Caspian Sea Reflected in the Multilateral Negotiations Among Caspian States.....	92
	6.5	New Regulations for the Use of Nonliving Resources in the Caspian Sea According to the Caspian Sea Convention 2018..	94
	6.6	Conclusion..	95
	References..		96
7	**The Legal Regime of the Living Resources of the Caspian Sea**..		99
	7.1	Tensions Between the Protection of Fish Stocks and the Oil Industry in the Caspian Sea........................	99
	7.2	Regime of the Living Resources in International Law......	101
	7.3	Historical Development of Regulations of Fishing in the Caspian Sea..	105
		7.3.1 Soviet–Iranian Fishery Regulation..............	105
		7.3.2 Agreement on the Conservation and Rational Use of the Aquatic Biological Resources of the Caspian Sea.................................	106
		7.3.3 Regulation on the Living Resources of the Caspian Sea According to Other Regional Agreements.....	108
	7.4	Regulation of the Living Resources in the Caspian Sea According to the Caspian Sea Convention of 2018.........	109
		7.4.1 Caspian Living Resources Regime in the Interstate Negotiations on the Convention on the Caspian Sea Legal Status..........................	110
		7.4.2 Regime for the Living Resources in the Caspian Sea Convention of 2018.....................	111
	7.5	Conclusion..	114
	References..		115
8	**The Legal Regime of the Pipelines in the Caspian Sea**.........		117
	8.1	Pipelines in the Caspian Sea.........................	117
	8.2	International Law on Pipelines.......................	120
	8.3	Regulations on Pipelines in the Caspian Sea as Reflected in the Multilateral Negotiations Among the Caspian Sea States Until 2018.................................	123

	8.4	Regulations of Pipelines in the Caspian Sea as Reflected in the Caspian Sea Convention of 2018..................	124
	8.5	Conclusion.......................................	125
	References...		126
9	**The Legal Regime of Maritime Navigation on the Caspian Sea**..		127
	9.1	Ship Navigation on the Caspian Sea....................	127
	9.2	The Legal Regime of Shipping in International Law........	129
	9.3	Historical Rules on Navigation in the Caspian Sea.........	134
	9.4	Navigation in the Interstate Negotiations According to the Draft Convention on the Caspian Sea Legal Status.........	135
	9.5	Navigation Regime According to the Convention on the Caspian Sea from 2018..............................	137
		9.5.1 Freedom of Navigation in the Caspian Sea.........	138
		9.5.2 Innocent Passage in the Caspian Sea.............	140
		9.5.3 Nonmerchant Vessels........................	142
		9.5.4 Right to Access the Ocean....................	143
	9.6	Conclusion.......................................	144
	References...		144
10	**Protection of the Marine Environment of the Caspian Sea**......		147
	10.1	Introduction......................................	147
	10.2	Protocols for the Enforcement of the Tehran Convention....	150
	10.3	Environmental Principles Applicable to the Caspian Sea....	151
		10.3.1 Principle of Sustainable Development............	151
		10.3.2 "Future Generations" Principle.................	153
		10.3.3 The Precautionary Principle...................	154
		10.3.4 "The Polluter Pays" Principle..................	155
	10.4	Prevention, Reduction, and Control of Pollution in the Caspian Sea.......................................	156
		10.4.1 Land-Based Pollution........................	157
		10.4.2 Pollution from Seabed Activities................	161
		10.4.3 Pollution from Other Human Activities...........	162
		10.4.4 Pollution by Dumping........................	163
		10.4.5 Pollution from Vessels.......................	165
		10.4.6 Environmental Emergencies...................	167
	10.5	Protection, Preservation, and Restoration of the Marine Environment......................................	168
		10.5.1 Protection of Biodiversity.....................	170
		10.5.2 Invasive Alien Species.......................	171
		10.5.3 Coastal Zone Management....................	174
		10.5.4 Fluctuation of the Caspian Sea Level.............	175
	10.6	Institutional Framework for Cooperation in the Legal Protection of the Caspian Environment.................	177
		10.6.1 Conference of the Parties.....................	177

		10.6.2	Secretariat	179
		10.6.3	Specific Institutional Arrangements	180
	10.7	Procedures		180
		10.7.1	Exchange of Information	181
		10.7.2	Monitoring	184
		10.7.3	Environmental Impact Assessment	187
		10.7.4	Reporting	191
		10.7.5	Consultations	193
		10.7.6	Public Access to Information	195
	10.8	Implementation of the Tehran Convention and Compliance		197
		10.8.1	Compliance	197
		10.8.2	Liability and Compensation	197
		10.8.3	Settlement of Disputes	199
	10.9	Conclusion		200
	References			201
11	**Concluding Remarks**			203
List of International Treaties				207
Bibliography				213

Abbreviations

AnnIDI	Institut De Droit International Annuaire
AIOC	Azerbaijani International Oil Consortium
ASEAN	Association of Southeast Asian Nations
BAT	Best available technology
BEP	Best environmental practice
BTC	Baku–Tbilisi–Ceyhan
CASPCOM	Coordinating Committee on Hydrometeorology and Pollution Monitoring of the Caspian Sea
CCWLF	Caspian Centre for Water Level Fluctuations
CEP	Caspian Environment Program
CIRM	Comite' International Radio-Maritime
CIS	Commonwealth of Independent States
CMI	Comite' Maritime International
CNPC	China National Petroleum Corporation
CPC	Caspian Pipeline Consortium
CPIT	China–Pakistan–Iran–Turk
CPUE	Catch-per-unit-effort
EC	European Community
ECSC	European Coal and Steel Community
EEC	European Economic Community
EEZ	Exclusive economic zone
EIA	Environmental impact assessment
EPIL	Encyclopedia of Public International Law
EQO	Environmental quality objectives
EU	European Union
EURATOM	European Atomic Energy Community
FAO	Food and Agriculture Organization
GA Res.	General Assembly Resolution
GATT	General Agreement on Tariffs and Trade
GEF	Global Environment Facility

IALA	International Association of Marine Aids to Navigation and Lighthouse Authorities
ICARCS	International Commission on Aquatic Resources of the Caspian Sea
ICJ	International Court of Justice
ICJ Rep.	International Court of Justice Report
ICLQ	International and Comparative Law Quarterly
IHO	International Hydrographic Organization
ILA	International Law Association
ILC	International Law Commission
ILEC	International Lake Environmental Committee
ILM	International Legal Materials
IMO	International Maritime Organizations
IMCO	Inter-Governmental Maritime Consultative Organization
ISO	International Organization for Standardization
ITCAMP	Integrated Transboundary Coastal Area Management Protocol
LBS	Land-Based Sources Protocol
LNTS	League of Nations Treaty Series
MARPOL	Marine pollution
MEPC	Marine Environment Protection Committee
NAFTA	North American Free Trade Agreement
NATO	North Atlantic Treaty Organization
NCAP	National Caspian Action Plan
NEAFC	North-East Atlantic Fisheries
NEAP	National Environmental Action Plan
NIOC	Commission National Iranian Oil Company
NJW	Neue Juristische Wochenschrift
OECD	Organisation for Economic Cooperation and Development
OJ	Official Journal of the European Union (since 1968), C series (Communications), L series (Laws)
RIAA	Reports of International Arbitral Awards
RSFSR	Russian Soviet Federative Socialist Republic
CACP	Central Asia-China Pipeline
SAP	Strategic Action Program for the Caspian Sea from 2003
SC Res.	Security Council Resolution
SCP	South Caucasus Pipeline
SOLAS	Safety of Life at Sea
TAC	Total allowable catch
TACIS	Technical Assistance to the Commonwealth of Independent States
TANAP	Trans Anatolian Pipeline
TAP	Trans Adriatic Pipeline
TAPI	Turkmenistan–Afghanistan–Pakistan–India
TDA	Transboundary diagnostic analysis
TITR	Trans-Caspian International Transport Route
TRACECA	Transport Corridor Europe–Caucasus–Asia

Abbreviations

UNCED	United Nations Conference on Environment and Development
UNCLOS	United Nations Convention ion the Law of the Sea
UNCTAD	United Nations Conference on Trade and Development
UN Doc.	United Nations documents
UNDP	United Nations Development Program
UNECE	United Nations Economic Commission for Europe
UNEP	United Nations Environmental Programme
UNTS	United Nations Treaty Series
US	United States
USSR	Union of Soviet Socialist Republics
WB	World Bank
WCED	World Commission on Environment and Development
WHO	World Health Organization
WMO	World Meteorological Organization
WTO	World Trade Organization

Chapter 1
Introduction and Course of the Investigation

1.1 Situation in the Caspian Sea After the Collapse of the Soviet Union

Is it possible for the Caspian Sea, which after the collapse of the Soviet Union in 1991 has become a bone of contention between the five bordering countries Russia, Kazakhstan, Turkmenistan, Azerbaijan, and Iran, to turn into an area of---literally speaking---fruitful cooperation? This question remains open for the time being, but there is a clear reason to be hopeful, considering the recent milestone in the legal cooperation of the riparian states. Contested since the dissolution of the Soviet Union, the legal status of the Caspian Sea now seems to have been defined. After almost 30 years of negotiations among five Caspian Sea states, which emerged in the course of the dissolution of the Soviet Union, a mutual consent over the problem over the international legal status of the Caspian Sea has been reached. On August 12, 2018, all riparian states have signed the Convention on the Legal Status of the Caspian Sea (hereinafter Caspian Sea Convention) during the fifth Caspian Summit in Aktau. This document shall serve as basis for conducting territorial demarcation and for the clarification of uncertainties about the extent of sovereign rights of the coastal states to the exploitation of Caspian natural resources. Also, the extent of shipping and fishing rights of states as well as the scope of rights on fossil resources have been settled. It gives a legal basis for the hope that previous obstacles for the economic development of the region, deteriorating environmental conditions, destabilizing of the political situation, and, as result, the lack of security in the Caspian region will step by step decrease.

Many years of ineffective attempts to define the legal framework of the Caspian Sea could be explained by pointing to the existing deep differences between the geopolitical and economic interests of the five littoral states. Along with the change of the geopolitical situation in the region after the collapse of the Soviet Union and the region's opening to international collaboration, mainly in the field of oil and gas resources, the Caspian Sea region has come to the center of attention for China, the

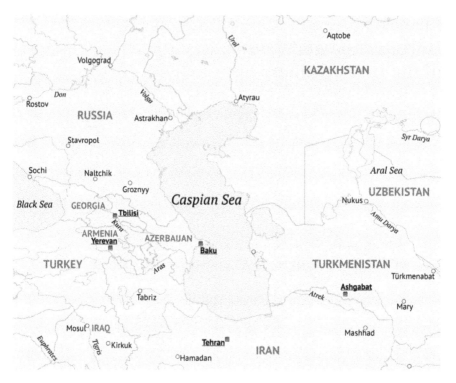

Fig. 1.1 Map of the Caspian Sea, 2018 Source: Wikimedia Commons, the free media repository https://commons.wikimedia.org/wiki/File:CaspianSeaDrainage_v1.png

US, and the EU, which has intensified the competition of powers existing in the region already for a long time. The signature of the Convention on the Legal Status of the Caspian Sea offers a starting point for the process in defining and implementing the rights and obligations of the riparian states with respect to the use of the Caspian Sea, including its waters, seabed, subsoil, as well as air space above (Fig. 1.1).

1.2 Adoption of the Caspian Sea Convention Defines the Importance of Research

The efforts to determinate the legal status of the Caspian Sea after the collapse of the Soviet Union followed in three parallel tracks. The main goal of the mutual negotiations conducted by all coastal states was the adoption of the Convention regulating the overall legal status of the Caspian Sea. The draft of this Convention was developed by the commission comprising vice ministers of foreign affairs and discussed until the end of the 1990s to be finally adopted in 2018. Simultaneously, between 1998 and 2003, the North Caspian coastal states prepared and signed bilateral agreements regarding the division of the seabed of the Caspian Sea into sectors for the exploitation of natural resources there (hereinafter North Caspian Agreements). The adoption of the North Caspian Agreements between Azerbaijan, Kazakhstan, and Russia, and a few years later also Turkmenistan, gave rise to doubts as to their legality among those Caspian littoral states, which have not been involved in their preparation and signature, mainly Iran. Today, we see, however, that the regulation of the North Caspian Agreements has been accepted by the newly adopted Caspian Sea Convention. The third parallel track of intergovernmental negotiations over the Caspian Sea was reflected in the ``step-by-step'' approach to the development of regulations concerning environmental protection issues and regional security. The Framework Convention for the Protection of the Marine Environment of the Caspian Sea, in 2013, was signed (hereinafter Tehran Convention) to regulate the issue of the fragile Caspian Sea environment threatened by the multiple economic activities undertaken by the coastal states in the Caspian Sea and, in 2014, the Agreement on the conservation and rational use of aquatic biological resources of the Caspian Sea. Additionally, the Agreement on cooperation in the field of prevention and liquidation of emergency situations in the Caspian Sea entered into force in 2017. In 2010, the Agreement on Security Cooperation in the Caspian Sea (hereinafter Caspian Security Agreement) was adopted to facilitate interstate cooperation in fighting terrorism, trafficking, and illegal exploitation of natural resources.

The multiple and often parallely undertaken efforts to regulate the use and protection of the Caspian Sea might have delayed the process of determination of the overall legal regulation but have finally not prevented the adoption of the Caspian Sea Convention, which crowns these efforts to establish a suitable framework for the lawful development and protection of the natural resources of the Caspian Sea.

Natural resources of a transboundary nature, like those in the Caspian Sea, are always the object of interest of all riparian states and therefore shall be based on international law.[1] Such an approach is reflected also in the global agenda for sustainable development and its 17 Sustainable Development Goals, which include clear targets for global and national policies to ensure the sustainable development of natural resources.[2] The legal status of a maritime territory, which bears such natural resources, as well as the clarification of legal regimes regulating the use of the resources are key for guaranteeing the sustainable development of the region. The newly adopted Caspian Sea Convention defines the scope of the states' sovereignty in the area, as well as their rights to use water, subsoil, resources, and regulates transit and transportation on the water, as well as the laying of underwater pipelines and cables. The Convention reinforces the regulations for the environmental protection of the Caspian Sea as well as regulates the security issues, challenges, and

---

[1] ILA Resolution, 3/2002, ILA Report of the Seventieth Conference, New Delhi.
[2] Transforming our world: the 2030 Agenda for Sustainable Development, Oct. 21, 2015, UN General Assembly.

threats in the region. All this confirms the importance of the investigation on the legal details concerning the new developments within the legal framework for the Caspian Sea shedding light on the current legal situation in the region.

1.3 Present State of Research Into the Legal Status and Regime of the Caspian Sea

The new edition of the book will discuss the recently adopted Convention on the Legal Status of the Caspian Sea from 2018. In the first edition of this book, an analysis of the legal status of the Caspian Sea was made based on the Draft of the Caspian Sea Convention, which was prepared and discussed during the years of multilateral negotiations by the coastal states undergoing until 2018. The current second edition of the book significantly differs from its previous edition. It reflects not anymore on the process but on the finally adopted document of the Caspian Sea Convention, which offers currently binding framework for the Caspian Sea and was adopted after almost 30 years of negotiations, which took place in the context of a significantly changed geopolitical situation around the Caspian Sea and the emergence of five independent states around its waters. The adoption of the Caspian Sea Convention offers a milestone in the understanding of the current state positions over the legal framework for the Caspian Sea. It gives a firm basis for the lawful development and use of natural resources as well as reinforces the principles for its protection. The Caspian Sea Convention offers a new regulation for territorial delimitation as well as maritime transport, fisheries, and other economic uses of the Caspian waters, its seabed, and its subsoils. The analysis of the provisions of the newly adopted Caspian Sea Convention represents the scientifically innovative part of this book. Since the adoption of the Caspian Sea Convention, there has been no comprehensive analysis of this new legal status regime of the Caspian Sea. Some publications undertook an analysis of the regional and local levels of newly introduced regulations, but these works have not comprehensively analyzed the whole texts of the Convention. Also, in the past two decades, there had been a number of specialized literature on this topic, publications on local[3] and international legal aspects concerning the Caspian Sea,[4] which were mainly limited to a discussion on the legal status of the Caspian Sea based merely on the legal framework developed under the USSR and Persia/Iran framework without comprehensively analyzing the outcomes of ongoing negotiations between the coastal states on the legal status of the Caspian Sea. A reading of a number of sources could offer an impression that some authors merely presented the political positions of each coastal states in the dispute

---

[3]See: (Barsegov 1998; Kolodkin 2002; Mamedov 2001; Merzlyakov 1998; Ranjbar 2004, and Salimgerei 2003).
[4]See: (Buttler 1971; Chufrin 2001; Elferink 1998a, b; Oxman 1996; Romano 2000; Uibopuu 1995, and Vinogradov and Wouters 1995).

over the status of the Caspian Sea rather than following the scientific impartiality required for a legal analysis of the subject. The aforementioned lack of literature was mainly due to the practical difficulty to access legal documents that reflected the status of interstate negotiations over the Caspian Sea. The newly adopted Caspian Sea Convention is nowadays publicly available and offers a ground for further scientific research and legal analysis, as presented in this edition of the book. The growing international interest in the Caspian region, mainly in the exploitation of the natural resources of the Caspian Sea, provides continues justification for research on the legal status and regime of the Caspian Sea.

The updated edition of this book offers insights into other interstate agreements concluded between the Caspian Sea states after 2014, when the first edition of the book was published, including the bilateral agreement between Kazakhstan and Turkmenistan on the use of natural resources in their relevant sectors of the Caspian Sea seabed and subsoil, as well as multilateral arrangements, including the Agreement on the conservation and rational use of aquatic biological resources of the Caspian Sea and a set of additional Protocols to the Caspian Security Agreement of 2010, as well as additional Protocols to the Tehran Convention of 2006 regulating the environmental conditions of the region.

1.4 Structure of the Book

This chapter is offering a general introduction to the content of the book and presenting the course of investigation. As first, it trusses the complex situation within the Caspian Sea region apart upon the collapse of the Soviet Union, which set the framework for the interstates negotiations of the Caspian Sea riparian states to define the new Legal Status of the Caspian Sea. The recent developments in the region and especially signature of the Convention of the Legal Status of the Caspian Sea explain the novelty and need for an update. The chapter highlights also the current state of the research over the legal framework for developing the cooperation of countries, and also it presents the structure of the book.

The Chap. 2 of this book offers a brief introduction into the changing geographical, political and economic conditions in the Caspian Sea region. It offers a broader regional picture to allow for a better understanding of the importance of the legal framework and regime, which has been newly adopt for the Caspian Sea. The clarity and transparency in the management of transboundary natural resources, as well as protection of the environment of shared water basins are a sine qua non condition for sustainable economic development as well as for political stability in the region. This chapter presents the situation in the Caspian Sea region after the collapse of the Soviet Union, as well as provides the overview of the current outcomes of decades of interstate negotiations over the legal status and regimes of use of the Caspian Sea, which ended up finally in the adoption of the Convention on the Caspian Sea Legal Status in 2018.

In Chap. 3, the historical overview of the legal framework applicable to the Caspian Sea use and protection will be presented. The agreements concluded between the USSR and Persia/Iran remained in force despite the collapse of the USSR and the emergence of the new independent countries for the last almost 30 years and served as a main basis for the assessment of the states' rights and obligations over this area and its resources. Based on these agreements, in the course of the development of the new legal framework for the Caspian Sea, there were heated debates on whether the Caspian Sea should be recognized in legal terms as a lake, sea, or condominium. The analysis of the legal consequences of such a qualification is briefly presented in this chapter. It revealed that the negotiating states left behind the legal-theoretical dilemma over whether the Caspian Sea is in a legal sense a sea or a lake and have developed a fairly complex set of provisions reflecting a unique legal system applicable to the Caspian Sea.

Further on, in Chap. 4, the book reflects on the states' position in the intergovernmental negotiations ongoing since the late 1990s. Three parallel tracks of these negotiations were presented in this chapter: firstly, the multilateral five-party negotiations on the development of the Caspian Sea Convention; secondly, the development of the North Caspian Agreements aimed at the use of natural resources between Russia, Kazakhstan, and Azerbaijan, and later Turkmenistan; and thirdly, the ``step-by-step" development of the multilateral agreements conducted by all five littoral states on the separate regimes of the Caspian ecosystem, as well as interstate cooperation on regional security. This chapter also discusses the separately adopted legal agreements regarding the security of the Caspian Sea, as well as the hydrometeorological conditions, and conservation and rational use of aquatic biological resources of the Caspian Sea.

Starting from the following chapter, a legal analysis of the separate aspects of the use and protection of the Caspian Sea was provided in reference to the provisions of the newly adopted Caspian Sea Convention.

The Chap. 5 discusses the delimitation of the Caspian Sea, which used to be always the most contested issue in settling the legal status of the Caspian Sea. This Chapter discusses the interrelations between territorial delimitation and the regime of the use of the Caspian Sea. It presents firstly the nonlegal aspects of settlement of the seaward boundaries of the Caspian Sea and follows with an elaboration on territorial delimitation and the issue of state sovereignty in the Caspian Sea before and after the adoption of the Caspian Sea Convention in 2018. The regulations of the bilateral and trilateral agreements on sharing the Northern Part of the Caspian Sea seabed and its subsoil for the use of its resources were also discussed. These regulations stay in force despite the adoption of the new Caspian Sea Convention and provide more detailed provisions for the management of the Caspian Sea of the Northern Part.

It's followed by analysis of the new Caspian Sea Convention, which has introduced the new maritime boarder lines in the Caspian Sea allowing for the defining of the scope of the states' territorial sovereignty, including the issue of living and nonliving natural resources. Highlighting the importance of territorial delimitation in the context of state sovereignty, the chapter offers and overviews over the state practice in delimitation of the Caspian Sea until 2018. It is offering a

1.4 Structure of the Book

brief overview of the bilateral and trilateral agreements adopted early 2000 between Kazakhstan, Azerbaijan and Russia (between 1998 and 2003), so called "North Caspian Agreement"[5] as well as finally with Turkmenistan (2014).[6] The new maritime zones in the Caspian Sea, despite steering clear of the law of the sea regulations, refer to the globally recognized legal models offering legal solutions for regional state conflicting interests while determining boundaries.

Chapter 6 reflects on the regime for the use of nonliving resources in the Caspian Sea---especially oil and gas---which significantly impacts states' economic development and which was one of the most acute reasons for riparian states to claim the delimitation of the Caspian Sea. First, this chapter provides information on the existing reserves of nonliving resources in the Caspian Sea. It discusses the existing international regulation of nonliving resources as a reference point for analysis of the newly adopted provisions of the Caspian Sea Convention. Additionally, it discusses the content of bi- and trilateral North Caspian Agreements, which were adopted in parallel with the overall negotiations on the legal status of the Caspian Sea from 1998 to 2003 and expended by a Kazakhstan-Turkmenistan agreement from 2014 and which remain in force.

Chapter 7 on the legal regime of the living resources of the Caspian Sea starts from describing continuous tensions between the protection of fish stocks and the oil industry in the Caspian Sea. It focuses on the presentation of the newly adopted regime for the regulation of the living resources of the Caspian Sea in comparison with the existing regime for living resources in the international law. The transition from the regime of common use of fish stock during the Soviet era to the new stage of legal regulation of living resources resulted in an extensive exploitation of these resources. The International Commission on Aquatic Resources of the Caspian Sea (ICARCS), the Tehran Convention and Agreement on the Conservation and Rational Use of Aquatic Biological Resources of the Caspian Sea, as well as the newly adopted Caspian Sea Convention, presented in this chapter, offer a comprehensive framework for the regulation and protection of the living resources of the Caspian Sea.

Chapter 8 deals with the legal regime of the pipelines in the Caspian Sea, showing the current framework for the use of pipelines in this area, these being an important way of transporting Caspian oil and gas resources towards the world markets. The chapter presents an overview of the international law on pipelines as a reference point for discussing regulations of the newly adopted Caspian Sea Convention. In the past, the legal regime of the Caspian maritime pipelines has never been subject to interstate agreements but rather has been an issue of general practice of the Caspian Sea states. Therefore, it is of special importance to analyze the Caspian Sea Convention, which regulated the pipeline regime and allowed the coastal states to lay transboundary pipelines and cables.

---

[5]Charney and Alexander (2003), pp. 1057 et seq. No. 5–12.

[6]Legal status of the Caspian Sea, Ministry of Foreign Affairs of the Republic of Kazakhstan https://www.gov.kz/memleket/entities/mfa/press/article/details/591?lang=en. Accessed 7 July 2020.

Chapter 9 discusses the legal regime of maritime navigation on the Caspian Sea, it being traditionally the most important regime for the use of the Caspian Sea. It presents the legal regime of shipping in international law as well as newly adopted regulations on shipping included in the Caspian Sea Convention in 2018. An initial regime of freedom of shipping was to some extent recognized as valid; however, some unique new regulations on the scope of shipping rights in the Caspian Sea, especially of third states, differentiate it from the standards existing in the international law of the sea. The chapter discusses freedom of navigation in the Caspian Sea including innocent passage regulation as well as write to access to oceans through the Volga River.

Chapter 10 elaborates on the protection of the marine environment of the Caspian Sea. It begins with an analysis of the environmental principles applicable to the Caspian Sea. Following the structure of the most comprehensive legal act providing for the protection of the Caspian environment---the Tehran Convention and its ancillary Protocols---this chapter adds the newly adopted environmental regulations, as settled in the Caspian Sea Convention, which has reinforced the importance of the provisions included in the Tehran Convention and existing environmental regulations. This chapter has been divided into parts concerning the prevention, reduction, and control of pollution of the Caspian Sea and the protection, preservation, and restoration of the marine environment. Further, it describes the institutional framework for cooperation in the legal protection of the Caspian environment and existing environmental procedures. With the adoption of the Tehran Convention, the states parties set specific environmental goals but avoided taking on explicit commitments. Its full legally binding effect can only be achieved through the adoption of implementing protocols, which takes place only gradually.

This book does not present the national legislation of the Caspian Sea littoral states concerned, although it is locally referenced. The main reason for this comes from the subject of the research, which is restricted to the international legal aspects of the regulation of the state and the regime of the Caspian Sea. Any provisions of national law can have no binding effect on the legal status of third countries and thus cannot contain requirements for these third countries. Since the adoption of the Convention on the Legal Status of the Caspian Sea, which happened only recently, it will take time to adopt the respective national regulations introducing its provisions into national legislation. This does not mean that the significance of coastal states' national legislation on the use of the Caspian Sea and its resources as well as its protection were put in question. Quite the contrary, awareness of the importance of national legal solutions and their complexity, which cannot be fully analyzed within this book, encourages only a local reference to the existing provisions, with the intention to avoid only a cursory presentation of complex national legal regulations.

An analysis on the legal status and regime of the Caspian Sea presented in this publication covers a period ending on July 1, 2020.

References

Anon (3/2002) ILA Resolution. London, ILA Report of the Seventieth Conference, New Delhi

Anon (2015) Transforming our world: the 2030 Agenda for Sustainable Development. UN General Assembly, October 21 (s.l.)

Barsegov J (1998) Kaspij v mezhdunarodnom prave i mirovoi politike (Caspian Sea in the international law and global policy). Institute of International Politics and International Relations, Russian Academy of Sciences, Moscow

Buttler W (1971) The Soviet Union and the law of the sea. John Hopkins Press, Baltimore

Charney J, Alexander L (eds) (2003) International maritime boundaries. Martinus Nijhoff, Dordrecht

Chufrin G (2001) The security of the Caspian Sea region. Oxford University Press, Oxford

Elferink AO (1998a) Denmark/Iceland/Norway-bilateral agreements on the delimitation of the continental shelf and Fishery Zones, 13th edn. IJMCL

Elferink AO (1998b) The legal regime of the Caspian Sea. Are the Russian arguments valid?. In: Wolfrum R (ed) The legal foundation of the New Russia. Norwegian Institute of International Affairs, Oslo

Kolodkin A (2002) Ne zhdat' u morya pogody (Don't whistle for a wind). Izvestya

Mamedov R (2001) International legal satus of the Caspian Sea; issues of theory and practice. Turkish Yearb Int Relat 32:217--259

Merzlyakov Y (1998) Legal status of the Caspian Sea // International life. M. (Pravovoy status Kaspiyskogo morya // Mezhdunarodnaya zhizn'). - N 11/12. - p. 135

Oxman B (1996) Caspian Sea or lake: what difference does it make?. Caspian Crossroads Mag 1 (4):14

Ranjbar R (2004) Das Rechtsregime des Kaspischen Meeres und die Praxis der Anrainerstaaten. Nomos, Baden-Baden

Romano C (2000) The Caspian Sea and international law: like oil and water. In: Mirovitskaya N, Ascher W (eds) The Caspian Sea: a quest for environmental security. Kluwer, Dordrecht

Salimgerei A (2003) Pravovoi Status Kaspisjskovo Moria (Legal status of the Caspian Sea). Kazakh State University Publishing House: Kazakh State University Publishing House

Uibopuu H (1995) Das Kaspische Meer und das Völkerrecht.. Recht in Ost und West. Zeitschrift für Ostrecht und Rechtsvergleichung 39(7):201

Vinogradov S, Wouters P (1995) The Caspian Sea: current legal problems. Zeitschrift für ausländisches öffentliches Recht und Völkerrecht 55:604--623

Chapter 2
Geography, Politics, and Economy in the Caspian Region

The following description presents the geographical, economic, and political aspects of the development of the Caspian region as the question of the legal status and regime of the Caspian Sea is going to be explained against that background.

Until just three decades ago, neither Europe nor America nor China were interested in the Caspian Sea, located in the barren steppes of Central Asia. Only its exceptional geographical size ensured it with the continued interest of scientists. The Caspian Sea is the largest enclosed inland water area on earth.[1] It is bordered by Azerbaijan, Iran, Kazakhstan, Russia, and Turkmenistan. While the large northern part of the Caspian is only about 6 m deep on average, its deepest point, located in the south, is 995 m below water level. The water masses of the Caspian Sea are fed by the rivers Volga, Ural, and Kura. The Caspian Sea has no natural connection to the oceans today, but there is a navigable link to the Sea of Azov and thus the Black Sea, the Mediterranean Sea, and the Atlantic Ocean[2] through the Volga, the Volga–Don Canal, and the Don River.

With the dissolution of the Soviet Union and the discovery of new oil and gas fields in the Caspian Sea, this region has become interesting to European countries as a potential source of raw materials. Also, China[3] and Japan[4] have expressed their economic interests in the region. The interest of the USA has been rather about expanding their sphere of influence in the post-Soviet space and pushing back Russia and Iran from the region. In this way, the entire Central Asia and the Caspian Sea have become a chessboard of world powers and thus got into the focus of

---

[1] The Caspian Sea is longer than 1200 km north to south and 300 km east to west on average (max 500 km). Its surface area is estimated at 436 km², but its water level is fluctuating. See: Brockhaus Enyzklopädie (2006), p. 650.
[2] See: (Rabinowitz et al. 2004).
[3] Kenderdine (2018).
[4] JOGMEC (2018).

international attention.[5] In addition, the attack on the World Trade Center on September 11, 2001, moved northwestern Central Asia, next to Afghanistan, into the public view. Thus, the region of the Caspian Sea has become a geopolitically sensitive area, not only due to the issue of fighting terrorism but primarily because of access to energy resources.

The modern conflict around the Caspian resources has its origin in the so-called Contract of the Century of 1994, when the newly formed Azerbaijan International Operating Company (AIOC) consortium of the Azerbaijani government signed—against opposition from the other coastal states—an agreement with western oil companies for 7.4 billion US dollars[6] for the development of known oil fields. Income from the existing oil and gas resources represents a major part of the gross domestic product of the newly formed Caspian states and thus also forms an indispensable basis for the independent political existence of those states, as to their further economic development.[7] Therefore, Azerbaijan, Kazakhstan, and Turkmenistan have been calling for a final delimitation of the Caspian Sea and thus its legal division. Russia and Iran, however, wanted rather to maintain control over the region, as they have for centuries. Therefore, Russia was trying to avoid the loss of control over the oil and gas pipelines, which was the main way of exporting Caspian resources, and to prevent the construction of new pipelines.[8]

In the geopolitical context, the Caspian-Central Asian region has by no means the energy implications of the Persian Gulf and plays a smaller role in global energy security. The reasons for it are as follows: firstly, by 2018, oil production in the littoral states of the Caspian Sea reached 20.6% of the world's production.[9] At the end of 2018, Caspian Sea littoral states had 2,994,000 barrels of proved reserves, which represents 17% of the world's total proved reserves.[10] Other estimates of the Caspian reserves originating from the 1990s could be rather assessed as a politically motivated overestimation.

Secondly, the transport of energy resources from the Caspian Sea to the world markets is proving to be challenging.[11] The development of additional oil and gas reserves in the Caspian region presupposes, however, that a large part of the energy resources is to be exported to international markets. For this, a strengthening of the existing infrastructure is needed. So far, exports have been handled over long overland routes, which might involve high costs (e.g., the railway transportation to Novorossiysk or to the Baltic ports). Chinese investments linked to the Belt and Road Initiative have brought major improvements along the main transport

---

[5]Taghizadeh-Hesary et al. (2019).
[6]Aliyev (n.d.).
[7]Aydin and Azhgaliyeva (2019).
[8]German (2014).
[9]BP (2019).
[10]See: (BP 2019).
[11]See: (Seck 1998).

corridors, including trans-Caspian shipping through Central Asia.[12] Some lines also pass via regions of long-standing political crisis (Chechnya). Due to the previous one-sided dependence on the Russian pipeline network, there is interest in building alternative transit routes mainly to China through Russia,[13] but many of them are still in the planning phase.

Thirdly, the littoral states of the Caspian Sea are politically unstable. The clash of the different interests and priorities of a number of countries has transformed the Caspian region into a politically sensitive area.[14] For centuries, the Caspian region had been characterized by the interaction of different ethnic, sociocultural, and religious traditions, which make today's geopolitical situation even more complicated. The conflicts that exist today can be broken down into the following groups: geopolitical, geoeconomic, ethnoterritorial, military strategic, environmental law, and religious conflicts.[15] An additional factor of political instability in the region is the never excluded possibility that elites of the neighboring states, west-oriented today, could be replaced overnight by ones that are hostile to the western powers.[16] Finally, the turbulent economic situation of the coastal countries blocked the way for a needed foreign investment in infrastructure in the Caspian region.[17] Luckily, it has been possible to avoid military conflict despite the unclear legal framework, but the point of conflicts has been sparking up frequently. There is hope that the adoption of the Caspian Sea Convention in 2018 will contribute to a lasting political stability and cooperation in the region.

So why is this remote region of such great strategic importance? Why did the United States count the Caspian region among its strategic national interests in the 1990s?[18] Obviously, there are other reasons than those previously mentioned, such as the main considerations of the international community, when it comes to the importance of the Caspian Sea for global energy security. The decisive factor seems to be the new balance of power that has developed. The contemporary struggle for oil and gas resources is sometimes referred to as the new Great Game, played in the nineteenth century by Russia and the United Kingdom. The Caspian region determines the future of US–Russian relations because, just like in Eastern Europe, the

---

[12]Martin Russell, "Connectivity in Central Asia, Reconnecting the Silk Road," EPRS | European Parliamentary Research Service, April 2019, accessed Feb 12, 2020, http://www.europarl.europa.eu/RegData/etudes/BRIE/2019/637891/EPRS_BRI(2019)637891_EN.pdf.

[13]Orazgaliyev (2017).

[14]See: (Kubicek 2013).

[15]See: (Zonn et al. 2010).

[16]Luke Coffey Time for a U.S. Strategy in the Caspian Aug 19, 2019 https://www.heritage.org/europe/report/time-us-strategy-the-caspian accessed June 12, 2020.

[17]Kushlis, Chris, and Ben Slay "Economic Overview of the Caspian Region" https://www.belfercenter.org/publication/caspian-conference-report-succession-and-long-term-stability-caspian-region accessed June 12, 2020.

[18]Elizabeth Jones "U.S. Caspian Energy Diplomacy: What Has Changed?" Caspian Studies Program, April 11, 2001 https://www.belfercenter.org/publication/us-caspian-energy-diplomacy-what-has-changed accessed June 12, 2020.

interests of the two former antagonists clash seriously in that region.[19] The oil factor has become a driving force[20] in Russian politics. The goal of the US seems to be to prevent Russia from restoring its former empire in the south of the Commonwealth of Independent States (further referred to as CIS).[21] One can refer to an example of pipelines, which besides economic reasons, are often determined by political motives. Similar is the case in the Caspian Sea, where constructing new pipelines can be seen as an attempt to control zones of influence[22] in the region. Central Asia is located as a connecting link in the heart of the Eurasian continent, and the system of land transport in the region could again become a revival of the historic idea silk route. Pipelines play an important role in raising the logistic position of the region and, in this context, seem to be of paramount importance. As an example, the political motivation is the US support for the idea of an "energy policy bridge" between Europe and the Caspian Sea and, thus, the emergence of new pipelines from Baku through Georgia to Ceyhan, a Turkish coastal town.[23] The project was concluded in 2005 despite strong opposition from Russia. Since then, Caspian oil could reach the western energy markets, bypassing Russia and Armenia, weakening Iran's position.[24] Further on, the US's attitude to Iran's involvement in the logistics of exporting raw materials from the Caspian Sea will impact Iran's desired position as the geoeconomic centre of the region.

In the 1990s, the European Union discovered its potential interests in the Caspian region. The EU's main motive for action was to ensure the diversification of its own energy supplies to avoid economic and political dependence.[25] Another reason was also the commitment of both Europe and the US, motivated by the intention to support and promote political stability in the Caspian region. On the one hand, the US and the EU tried to include the Caspian littoral states in the military cooperation with the North Atlantic Treaty Organization (NATO)[26] as part of the "Partnership for Peace" program and to move them closer to the Turkish sphere of cultural and economic influence. On the other hand, the EU provided financial support for the development of market economy in the post-Soviet republics of the Caspian region

---

[19]See: (Coffey 2019).

[20]See: (Pritchin 2019).

[21]Pritchin (2019).

[22]Kubicek (2013).

[23]Laurent Ruseckas, 'US Policy and Caspian Pipeline Politics: The Two Faces of Baku-Ceyhan', August 15, 2000 in http://ksgnotes1.harvard.edu/BCSIA/Library.nsf/pubs/ruseckas accessed July 13, 2020.

[24]Iran's Eastern strategy, Strategic Comments, (2018) 24:9, viii–ix, DOI: https://doi.org/10.1080/13567888.2018.1557845.

[25]Hasmik Khachatryan The Southern Gas Corridor-a guarantee to the EU's energy security?, 26 April 2019 http://aprei-project2018.com.ua/wp-content/uploads/2019/05/SGS-EU-energy-security-1.pdf accessed July 13, 2020.

[26]Frappi (2014).

through programs such as "TACIS" or "TRACECA,"[27] with the aim of supporting the democratization process in the region. The routes for the transportation of resources are of geostrategic importance.[28] While oil companies, for financial and security reasons, prefer the route through Georgia[29] and the Bosporus via the Trans Anatolian Pipeline (TANAP), Azerbaijan started delivering gas to Turkey in mid-2018 and was poised to send gas to Italy via the Trans Adriatic Pipeline (TAP) by 2020;[30] however, the construction of this pipeline in Italy has not yet been completed, so the gas delivery is delayed until almost the end (October) of 2020.[31]

Despite China entering the Great Game in the Caspian region later than the other global players, its energy cooperation with the Caspian littoral states has progressed significantly. Over the last decade, China has replaced Russia as the main export destination for Central Asian gas.[32] The role of the Caspian region will be significantly affected by the new actors, especially China[33] and Japan.[34] To them, the Caspian Sea is going to play a leading role in the future supply of strategic raw materials for Asia. For the new littoral states of the Caspian Sea, their countries' oil and gas resources are the only way of stopping economic decline. Oil has been a political weapon in the struggle for independence.

Due to various problems after the collapse of the Soviet Union and the correspondingly growing international economic as well as political interests in the region, the need for a solution to the question on the viable legal status of the Caspian Sea became inevitable. In 2018, finally after almost 30 years of negotiations, the Caspian Sea Conventions has been signed. Without a doubt, the adoption of the Caspian Sea Convention shall be recognized as a milestone in the negotiations over the states' cooperation of the Caspian Sea. The intergovernmental struggle to develop a legal status and regime for the development of the Caspian Sea's natural resources, its protection, and its transport has commenced, opening a new perspective for regional collaboration. The Caspian Sea waited a long time for the legal clarification of its problems until its coastal states finally managed to understand that

---

[27] Strategy of the IGC TRACECA 2016-2026, TRACECA official site http://www.traceca-org.org/en/home/strategy-of-igc-traceca-2016-2026/ accessed July 13, 2020.
[28] See: (Chufrin 2001).
[29] Bayramov (2019).
[30] Luke Coffey, Efgan Nifti "A Trans-Caspian Gas Pipeline: Start Small but Aim Big" May 16, 2019, National Interest https://nationalinterest.org/feature/trans-caspian-gas-pipeline-start-small-aim-big-58012 accessed July 13, 2020.
[31] Reuters, Nailia Bagirova: "Azeri gas for Europe from Shah Deniz II delayed to Oct 2020-TANAP head" November 1, 2019 See: https://www.reuters.com/article/azerbaijan-energy-gas/azeri-gas-for-europe-from-shah-deniz-ii-delayed-to-oct-2020-tanap-head-idUSL8N27G6C9 accessed July 5, 2020.
[32] Pirani (2019) and Guo et al. (2019).
[33] Chen and Fazilov (2018).
[34] See: (Pirani 2019).

legal clarity and stability always go along and are a key to comprehensive development in the region.

References

Aliyev I (n.d.) Oil and Gas Projects. [Online] Available at: https://en.president.az/azerbaijan/contract. Accessed 12 June 2020
Aydin U, Azhgaliyeva D (2019) Assessing Energy Security in the Caspian Region: The Geopolitical Implications for European Strategy. ADBI Working Paper (s.l.)
Bayramov A (2019) Great game visions and the reality of cooperation around post-Soviet transnational infrastructure projects in the Caspian Sea region. East Eur Polit 35(2):159–181
BP (2019) Statistical Review of World Energy 68th edition. pp 17 et seq. [Online] Available at: https://www.bp.com/content/dam/bp/business-sites/en/global/corporate/pdfs/energy-economics/statistical-review/bp-stats-review-2019-full-report.pdf. Accessed 18 July 2020
Brockhaus A (2006) Brockhaus Enzyklopädie, 21st edn. Bibliographisches Institut & F.A
Chen X, Fazilov F (2018) Re-centering Central Asia: China's "New Great Game" in the old Eurasian Heartland. Palgrave Commun 7:71
Chufrin G (2001) The security of the Caspian Sea region. Oxford University Press, Oxford
Coffey L (2019) Time for a U.S. Strategy in the Caspian (б.м.: б.н)
Frappi C (2014) The Caspian Sea Basin in United States strategic thinking and policies. In: Garibov A, Frappi C (eds) The Caspian Sea Chessboard. Milan, Egea, pp 181–202
German T (2014) Russia and the Caspian Sea: projecting power or competing for influence. United States Army War College Press, Carlisle Barras
Guo F, Huang C, Wu X (2019) Strategic analysis on the construction of new energy corridor China–Pakistan–Iran–Turkey Energy Reports. Energy Rep 5:828–841
JOGMEC (2018) Japan Oil, Gas and Metals National Corporation Provides Equity Financing to INPEX's Kashagan Oil Field Development in North Caspian Sea, Republic of Kazakhstan. [Online] Available at: http://www.jogmec.go.jp/english/news/release/news_06_000062.html. Accessed 12 June 2020
Kenderdine T (2018) Caucasus Trans-Caspian Trade Route to Open China Import Markets. [Online] Available at: https://www.eastasiaforum.org/2018/02/23/caucasus-trans-caspian-trade-route-to-open-china-import-markets/. Accessed 2 June 2020
Kubicek P (2013) Energy politics and geopolitical competition in the Caspian Basin. J Eurasian Stud 4(2):171–180
Orazgaliyev S (2017) Competition for pipeline export routes in the Caspian region: the new Great Game or the new Silk Road?. Cambridge J Eurasian Stud 1
Pirani S (2019) Central Asian Gas: prospects for the 2020s. Oxford Institute for Energy Studies
Pritchin S (2019) Russia's Caspian Policy. Russ Anal Dig 235:3 et seq
Rabinowitz P, Yusifov M, Arnoldi J, Hakim E (2004) Geology, oil and gas potential, pipelines and the geopolitics of the Caspian Sea region. Ocean Dev Int Law 35:19–40
Seck A (1998) Pipelines from Central Asia and the Transcaucasus: a maze of alternatives. In: Pratt MA, Schofield CH, Blake GH (eds) Boundaries and energy: problems and prospects. Kluwer Law International, London
Taghizadeh-Hesary F, Yoshino N, Chang Y, Rillo A (2019) Achieving Energy Security in Asia: Diversification, Integration and Policy Implications. ADBI (б.м.)
Zonn I, Kostianoy A, Kosarev A, Glantz M (2010) The Caspian Sea Encyclopedia, 1st edn. Springer, Berlin

Chapter 3
Legal History and the Present State of Use of the Caspian Resources

3.1 Introduction

The name of the "Caspian Sea" does not uncover its legal status of today. The name "Caspian" comes from one of the tribes of "Caspian" who once inhabited the west coast of the Caspian Sea. Previously, the Caspian Sea was alternately designated by nearly 40 different names, which either had an ethnic origin (e.g., in the Russian zone: Hvalinsk Sea, in the Persian zone: Hazar Sea) or carried the name of the coastal cities or states (Baku Sea, Girkan Sea, Abeskun Sea).[1] The term "sea" for the Caspian Sea does not have any legal reference. Ancient scientists and geographers such as Herodotus, Aristotle, and Eratosthenes labeled the Caspian Sea as a closed basin or an ocean bay. These traditional notions have clearly geopolitical but no legal significance.[2] In the nineteenth century, thanks to the conclusion of Russian–Persian treaties, a first reference regarding the legal status of the Caspian Sea was made. The provisions introduced after the First World War, which remain in force until today, have many omissions or are partly obsolete. Consequently, based on Soviet–Iranian agreements and regional customary law, the current legal principles governing the Caspian Sea either don't give a clear understanding of the legal status of the Caspian Sea or no longer appear sufficient to deal with new complex political, economic, and environmental problems. This would suggest the need for a new set of provisions regarding the Caspian's legal status.

The concept of legal status has to define the scope of a particular state's sovereign power over the water area in question. It defines how much rights and obligations and to what extent a state may exercise control over a relevant territory. It is important for proper comprehension of this analysis to draw a distinction between "legal status" and "legal regime."[3] In modern international law, both of these terms

---

[1] See: (Jiloe 1960).
[2] See: (Gull 1960).
[3] See: (Kolodkin 2002a).

are applied in the legal regulation of international lakes and seas. The classification of a pool of water into one of the two categories above—a sea or a lake—is of essential importance for determining both the status and the legal regime of the pool in accordance with, respectively, the law of the sea (for water pools qualifying as seas) or international water law (for pools qualifying as international lakes). Therefore, the term "legal regime" is defined in contrast with legal status as the particular set of rights and obligations of the states with respect to the use of the relevant area. The Framework Convention for the Protection of the Marine Environment of the Caspian Sea (Tehran Convention) is, for instance, classified as a regime issue. This chapter does not discuss issues relating to the legal regime of the Caspian Sea but concentrates on deliberations concerning the status of this water basin. The rules of public international law that are applicable to the Caspian's legal status depend primarily on the legal character of this body of water. How the Caspian Sea is regulated will therefore depend on its legal classification and the accompanying body of law. If the Caspian Sea is a "sea" in legal terms, the United Nations Convention on the Law of the Sea (UNCLOS) would be applicable. If, on the other hand, the Caspian Sea is a "lake" in legal terms, then customary international law concerning border lakes would apply. Given the great number of often contradictory legal opinions on its status, it seems that, rather than looking for new future regulation, it would be more rational to try to assess the compatibility of every option with current international public law doctrine. As the Caspian does not appear to fall into either category, it is therefore necessary to take into consideration its historical, geophysical, and legal peculiarity in deciding whether it is an international lake or an enclosed sea.[4] The weight of the here presented arguments suggests that the Caspian Sea is not governed by a condominium regime. It also does not appear to be a sea, so UNCLOS does not apply directly. Nor does it seem to be an international lake. So which legal concept should be applied to successfully define the legal status of the Caspian? It is clear that there are great difficulties in resolving this issue. At the current stage of interstate negotiations regarding the future legal situation of the Caspian Sea, as well as in the contemporary practice of states regarding the Caspian Sea, there is, however, absolutely no reference to either of the two categories mentioned above, which are today only referred to in research papers on that subject. What remains beyond dispute is that the Caspian Sea is a water basin surrounded by five sovereign states, which must determine its future legal status. After the benefits for each of these states are considered, such a decision needs to be mutual because only a multilateral solution can guarantee the legal stability of the Caspian region.

---

[4]Other legal concepts of classification of transboundary water basins apply as well. For instance, the Soviet–Iranian theory of closed-sea or the concept of the Condominium, which will be analyzed later on.

3.2 Russian–Persian Treaties Concluded Until the Nineteenth Century

Until the end of the twentieth century, the political arena of the Caspian region was alternately dominated by Persia and Russia. The Persian reign in the Caspian region began in the eighth century with the Abbasid dynasty and was challenged only after almost a thousand years by tsarist Russia, which was finally overthrown by Peter the Great in his first Persian campaign in 1722–1723.[5] At that time, the first agreements between Russia and Persia were concluded, which shall be seen as the beginning of the formation of the international legal status of the Caspian Sea. These treaties, however, included no reference to the use of the Caspian Sea and its resources.

In the Treaty of St. Petersburg of 12 September 1723, Persia lost Derbent, Mazandaran, Astarabad, and Baku to Russia, which thus practically enforced Russia's exclusive navigation rights on the Caspian Sea.[6] The subsequent cooperation Treaty of Rasht of 1729 on the demarcation and cession of certain territories, which provided for freedom of commerce and navigation, transferred to Persia its coastal areas previously conquered by Russia and allowed Persia once again direct access to the Caspian Sea. Further limitation on the influence of Persia in the Caspian Sea after two wars lost by Persia against Russia was reflected in two treaties concluded with Russia, the Treaty of Golestan of 1813 and the subsequent Treaty of Turkomanchai of 1828. They provided Russia with the exclusive right to have a naval fleet in the Caspian Sea.[7] Persia received merely rights of navigation in the Caspian Sea, and its commercial vessels were allowed to call at Russian port facilities. The treaties laid down the legal regime of navigation in the Caspian Sea for the first time. The use of natural resources, however, remained unregulated. The question of state borders between the coastal states was not clarified either, and thus the legal status of the Caspian Sea remained unresolved.

Both treaties remained in force until the period after the First World War. Attempts were made to indirectly replace the lack of explicit regulation of maritime borders within the Caspian Sea through the settlement and extension of land borders over the maritime areas. On December 9, 1881,[8] and May 27, 1893,[9] two treaties were signed between Persia and Russia, which determined the land borders between the two states eastward from the Caspian Sea. Article II of the first treaty stated that the exact line of the interstate border would be settled by special commissioners appointed by the parties. The treaty mentioned Astara and Hosseingholi as endpoints of the boundary lines on the shores of the Caspian Sea. Although the line between

---

[5]Mamedov (2001), pp. 109–114.
[6]Diplomatic Dictionary (1985), p. 483.
[7]Article 5 of the Golestan Treaty of 1813 and Article 8 of the Turkomanchai Treaty of 1828.
[8]Article 1 of the Convention between Persia and Russia of 1881.
[9]Article III of the Convention between Russia and Persia for the Territorial Interchange of Faruze in Khorassan, belonging to Persia, and Hissar, within the confines of the Transcaspian Region, and Abbas Abad, on the right bank of the River Araxes, belonging to Russia.

Astara and Hosseingholi served exclusively as a land border, some authors developed the idea that the extension of this line should be seen as a maritime boundary in the Caspian Sea between the two states.[10] However, as this concept was not of a contractual nature, it required a legal determination to reduce the potential of conflict between the coastal states. This happened only partially after the First World War, thanks to the conclusion of new agreements between the Soviet Union and Iran and the respective interstate practice.

3.3 Legal Heritage of the Twentieth Century in the Caspian Region

The Treaty on Friendship and Cooperation between the Russian Soviet Federative Socialist Republic (RSFSR) and Persia adopted on February 26, 1921 (further referred to as the 1921 Treaty) abolished their bilateral legal relations established in the nineteenth century. This agreement was the first one of a series of bilateral treaties completed between the USSR and Iran to regulate the use of the Caspian Sea. It remained one of the main sources of law in the Caspian region and bases for bilateral relations between the two states. The objective of the agreement of 1921 was the restoration of friendly relations between the two nations ("friendship and brotherhood," according to the original wording of the agreement). With the intention to abolish the hostile policy of the tsarist government against Persia, the treaty recognized existing borders and advocated for the avoidance of interference in the internal affairs of each other (Articles 2–4). Except for the restoration of Persia's equal rights of navigation, it did not specifically address the issue of the legal regime of the Caspian. Natural resources were mentioned only in connection with the renewal of fishery agreements. The treaty introduced the equality of both parties with regard to navigation in the Caspian Sea (Article 11), recognizing Persia's rights to keep warships in the Caspian Sea. In addition, the Persian government recognized the great importance of the Caspian fisheries as a food supply for Russia and respectively promised to conclude a new fishery treaty with the RSFSR (Article XIV). As a result of this commitment, a treaty regarding fishery on the southern Caspian coast was signed on October 1, 1927.[11] It stipulated that commercial fishing rights seaward of the coastal zone of 10 nautical miles would belong exclusively to a company 50% of which was owned by one of the parties (Article 5). This agreement was concluded for a period of 25 years and was not extended by Iran.

During the 1930s, the increasing navigation and fishing in the Caspian Sea resulted in bilateral negotiations to develop the existing legal framework. When it comes to navigational issues, the 1935 Treaty of Establishment, Commerce and

---

[10]See: (Polat 2002).

[11]Agreement regarding the Exploitation of the Fisheries on the Southern Shore of the Caspian Sea, with protocol, and Exchange of Notes.

Navigation between Iran and the Union of Soviet Socialist Republics (further referred to as 1935 Treaty) first replaced the Convention of Establishment, Commerce and Navigation of 27 October 1931[12] (further referred to as the 1931 Treaty), but in 1940, it was replaced by the Treaty of Commerce and Navigation between the USSR and Iran[13] (further referred to as the 1940 Treaty). The Treaty of 1940 was initially adopted with effectivity of three years, but because no party terminated it, the treaty entered into force for an unlimited period and retains its validity until today.

In both the 1935 and 1940 Treaties, states reserved navigation (military and merchant) as well as fishing for vessels flying their flags. They therefore excluded third states from the Caspian Sea and restricted the rights of innocent passage of ships of these other states. Nationals of third states were not even allowed to be crew members or port personnel (Article 139). Also, equal treatment of all vessels calling at, staying, and leaving ports as well as equal charges for services were guaranteed for ships flying the flag of a contracting party (Article 12). Both treaties provided for freedom to fish for both states in the entire Caspian Sea but within a 10-mile zone along their respective coasts, where each state had exclusive fishery rights. Beyond the fishing zone, both countries enjoyed unrestricted freedom of fishing for their residents.

Other activities, such as marine scientific research, were not mentioned in the 1940 Treaty. The coastal states did not resolve the issue of boundary lines in the Caspian Sea; thus, the question of territory covered by the national sovereignty of the littoral states, including over the use of natural resources in the Caspian Sea, remained unresolved. Oil and gas exploration and drilling in the areas adjacent to the coast were mentioned in the treaty in a highly unclear way. Iran agreed to grant the USSR "the right to set up petrol pumps in Iran and to construct petroleum storage depots and other buildings necessary for dealing in petroleum and its products" in conformity with existing laws and regulations in Iran (Article 9(8)). The common principle of the treaty is exclusivity of the rights of the coastal states regarding the use of the Caspian Sea, including its natural resources. This was based on the legal assumption that the Caspian Sea is a "Soviet–Iranian Sea," giving exclusive rights to shipping and fishing only to the USSR and Iran, as well as other usage of the Caspian Sea. The nonlittoral states in the Caspian region were refused any rights to the use of the Caspian Sea, which was repeatedly expressed in the official correspondence between the USSR and Persia/Iran.[14]

---

[12]In Article 16–17 it bars from traffic and fishing in the Caspian Sea all ships that do not fly the flag of the USSR or Iran.

[13]The term Persia was in use for centuries and was originally dedicated to the Persis (Pars or Parsa also, as modern Fars) which is a well region in southern Iran. However, the Persians themselves called their country "Iran," meaning "a land of Aryans." The name Iran was officially adopted in 1935. See: "Persia." Encyclopædia Britannica (2005).

[14]Exchange of Notes between the Persian Foreign Minister and the Soviet Ambassador, 27th October 1931, in: British and Foreign State Papers, volume 134, pp. 1045–1046; Exchange of Notes between the Soviet Ambassador at Tehran and the Iranian Minister for Foreign Affairs, 27th

With the adoption of bilateral treaties between the USSR and Persia/Iran in 1921 and 1940, a final basis for the rights and obligations of coastal states was settled on. As they were merely partially complete and regulated only the legal issues of shipping and fishing in the Caspian Sea, the lawful behavior of the littoral states was guided by local custom. After the collapse of the Soviet Union, the ambiguities regarding the legal status of the Caspian Sea caused a long-term legal dispute among the coastal states regarding the interpretation of these treaties as well as their binding status.

The long and fruitless legal debate of the 1990s over the framework of the legal status of the Caspian Sea regarded the main question whether the Soviet–Iranian treaties provide for the status of the Caspian Sea as a lake or as a sea in the legal sense and, thus, which of the international sets of principles—characteristic for an international lake or sea—should be applicable for the future status of the Caspian Sea. This issue, however, was completely disregarded in later practice of the coastal states, especially in light of the tendency to absorb the principles of applicable law of the sea and their inclusion into the Draft Caspian Sea Convention. Additionally, some states have concluded bilateral agreements concerning separate aspects of the use of the Caspian Sea, where the issue of whether it is a sea or a lake was not addressed.

The new geopolitical situation in the region after the collapse of the Soviet Union created a new legal situation, which opened a long-term dispute regarding the binding force of the Soviet–Iranian Treaties of 1921 and 1940 for the new littoral states of the Caspian Sea. Despite frequent denials of their legal binding force by the newly independent states, these agreements remain valid and shall be accepted as a basis for the interpretation of the legal status of the Caspian Sea, which will be shown in the following chapter.

3.4 State Succession and the Soviet–Iranian Agreements

With the creation of the Commonwealth of Independent States (CIS) according to its Founding Agreement of 8 December 1991 and the Alma-Ata Protocol[15] of 21 December 1991 signed by eleven former republics, the Soviet Union ceased to exist. However, questions regarding the succession of the newly independent states to the international treaties concluded by the USSR and their binding force for the successor states remained disputable for many years. A background to the discrepancies comes out of the fact that the nature of the succession of the former Soviet

---

August 1935, in: Soviet Treaty Series (1950), vol. II, pp. 145–146; Exchange of Notes between the Soviet Ambassador at Tehran and the Iranian Minister for Foreign Affairs, 25th March 1940 in British and Foreign State Papers, vol. 144, p. 431.

[15]ILM 31 (1992), Nr. 1, S. 147–154.

3.4 State Succession and the Soviet–Iranian Agreements

Union is controversial to scientists and politicians.[16] It is put forward by the former Soviet republics around the Caspian Sea—Azerbaijan, Turkmenistan, and Kazakhstan—that agreements concluded by the USSR, including the Soviet–Iranian Treaties of 1921 and 1940, lost their validity after the collapse of the Soviet Union. Therefore, the rights and obligations incorporated into these treaties were no longer binding for the newly established states. As a basis for this assertion, it is particularly emphasized that under international law, the USSR no longer exists as a contracting party.[17] The lack of their legal validity is also disputed because these treaties did not define boundaries between the former Soviet republics nor settled their legal status as a whole, referring merely to the issues of fishing and navigation.

Support for the abovementioned view would mean—something one can hardly agree with—that since the collapse of the Soviet Union, the Caspian Sea has been in a legal vacuum and needs an entirely new regulatory system. The legal consequences of states' succession are regulated by the Vienna Convention on Succession of States in Respect of Treaties, promulgated in 1978.[18] Although this Convention was not signed by any of the Caspian littoral states and neither is it recognized as part of customary international law by the scientific community, it still may constitute the main reference point for the solution of the question of succession of states in the Caspian region.[19] The Vienna Convention on the Law of Treaties, which was concluded in 1969[20] and which defines the conditions under which international agreements no longer apply, is not applicable to the dispute around the Caspian Sea. This agreement does not apply in cases of border treaties and therefore cannot be used for the assessment of the validity of the Soviet–Iranian Treaties of 1921 and 1940. These treaties, although they did not refer directly to state borders in the Caspian Sea, were aimed at delineating the spheres of influence of the neighboring states, intentionally leaving the boundaries in the Caspian Sea open.

The Vienna Convention on Succession of States in Respect of Treaties (Article 34 I a) states that "when a part or parts of the territory of a State separate to form one or more States, whether or not the predecessor State continues to exist, any treaty in force at the date of the succession of States in respect of the entire territory of the predecessor State continues in force in respect of each successor State so formed." This basic principle provides that the newly established states remain bound by the agreements of their predecessor. There are two exceptions to the general rule: first, when the states concerned agree otherwise (Article 34 II a) and, second, when it appears from the treaty or is otherwise established that the application of the treaty in

---

[16] As an example, the Baltic States may be mentioned, which are of the legal opinion they had been illegally occupied by the Soviet Union, but during the period of occupation they continued to exist *de jure* as subjects of international law. They could not be considered as successor states of the Soviet Union, see more detail: Schweisfurth (1992).

[17] Position of Kazakhstan on the legal status of the Caspian Sea, in: UN Doc. A/52/424, p. 3.

[18] ILM (1978), vol 17.

[19] See: (Ipsen 2004).

[20] Article 62 II a.

respect of the successor state would be incompatible with the object and purpose of the treaty or would radically change the conditions for its operation.

For the newly independent Caspian littoral states of Azerbaijan, Kazakhstan, and Turkmenistan, which were questioning the validity of the Soviet–Iranian treaties, this provision of the Vienna Convention on Succession of States in Respect of Treaties means that they remain bound by the former agreement of their predecessor state. Additionally, the newly independent states have expressed public statements explicitly reaffirming their consent to the treaties adopted by the Soviet Union. The CIS Founding Agreement of 8 December 1991 in its Article 12 includes a clear commitment of the newly independent states to fulfil the obligations deriving from the treaties and agreements concluded by the former Soviet Union. This statement regarding commitments toward the third countries was repeated in the Minsk Declaration[21] and in the Alma-Ata Declaration. The latter guaranteed "the discharge of the international obligations deriving from treaties and agreements concluded by the former Union of Soviet Socialist Republics."

With the collapse of the Soviet Union, there was some legal confusion as to whether this process shall have been assessed as secession—which implies that the predecessor state remains a subject of international law but experiences a changed territorial status—or a *dismembratio*—which implies a complete dissolution of the predecessor state and the creation of several new states on its territory. Russia itself claimed not to be a successor state of the USSR but to be its "continuator state."[22] Such a special status seemed to be confirmed by Russia overtaking the USSR's seat on the United Nations Security Council. Unlike the other former Soviet republics, only Russia did not need to receive recognition from third countries.[23] Some authors refer also to the fact that Russia was the only Soviet republic that did not make any declaration of independence, which shall be understood to mean that all republics that were split apart from the USSR were not considered as independent states, except Russia, which continues to be the USSR.[24] This argument is contradicted by the finding that the declarations of independence at that time did not mean separation from the USSR. Russia was also one of the states-parties to the CIS Founding Agreement, which terminated USSR's founding Treaty of 1922.[25] However, Russia never claimed legal continuity of the Soviet Union itself.[26] Therefore it is

---

[21]Europa-Archiv 1992, episode 8, p. D 302.

[22]e.g. Russian–British memorandum on consular missions of 30th January 1992, see Bulletin of International Treaties, 1993, no. 1, p. 33; Declaration on Russian–Japanese relations of 13th October 1993, see Bulletin of International Treaties, 1994, no. 2, p. 66. However, at the same time in other international legal writings it presents itself as a successor, see Unilateral acts of Russia: Government Decision of 11th March 1994, in: SAPP 1994, no. 12, pos. 983; International legal acts of Russia: The Protocol to the troop withdrawal agreement with Poland on 22nd May 1992, Bulletin of International Treaties, 1994, no. 2, p. 10.

[23]Ibid., pp. 175 et seq.

[24]See: (Antonowicz 1991–1992).

[25]See: (Schweisfurth 1992), pp. 172–173.

[26]See: (Schweisfurth 1996).

to say that the division of the USSR was a *dismembratio* and all CIS countries, including Russia, are successor states of the USSR. Thus, there is no subject identity between the former Soviet Union and today's Russian Federation.

The renunciation of the existing, though still incomplete, legal status of the Caspian Sea is linked to the question of the newly independent states' legal succession under the Vienna Convention on Succession of States in Respect of Treaties. Thus, in the Caspian Sea case, the rights and obligations of the predecessor state—i.e., the former Soviet Union—and its successors arising from international legal acts—including the Treaty of 1921 and Treaty of 1940—are equally binding on both.

3.5 Legal Interpretation of the Soviet–Iranian Treaties

Recognition of the validity of the Soviet–Iranian treaties for the newly independent states of Azerbaijan, Kazakhstan, Russia, and Turkmenistan is an important but insufficient assumption for defining the current legal situation in the Caspian Sea. Incompleteness of the Treaty of 1921 as well as of the Treaty of 1940 caused an inconsistent interpretation of these treaties by the newly established Caspian nation states. It led to the debate over the legality of the measures taken by states in the Caspian Sea.

In 1991, after the disintegration of the USSR, Azerbaijan, Kazakhstan, and Turkmenistan—three newly independent Caspian states—challenged the legal validity of the Caspian treaties,[27] which remained uncontested legally, either by the international community or by any of the signatory states, for several decades. The renunciation of the existing, though still incomplete, legal regime of the Caspian Sea was linked to the diverging negotiating positions adopted by the Caspian coastal states because of the historical-legal ambiguities of the status of the Caspian Sea as a sea, lake, or condominium. An official adoption of an unequivocal interpretation of the legal character of this body of water—whether as a sea, lake, or condominium—by the negotiating states would have had a serious impact on the scope of rights and obligations of the Caspian littoral states, also in respect of the Caspian resources. It would have become the basis for the legal interpretation of the existing Soviet–Iranian Treaties of 1921 and 1940. The scope of rights and obligations recognized in such a way would remain valid until the Soviet–Iranian treaties would be replaced by new rules. A monitoring of the development of the political debate suggested, however, that the Caspian did not appear to fall into either category. One could even argue that after the 1990s the question of legal classification of the Caspian Sea as a "lake" or "sea" was not debated in the intergovernmental negotiations, aiming at development of the future Convention on the legal status of the Caspian Sea. The outdated character of this approach is evident in the conclusion of treaties separately

---

[27] See: (Vylegjanin 2000).

regulating single regimes for the use of the Caspian Sea. Nevertheless, a cursory overview of this legal debate of the 1990s appears to require a full picture of the development of the legal relations in the Caspian Sea.

3.5.1 Caspian Sea as a "Sea"

The concept of the Caspian Sea being a sea in a legal sense can be traced back to the state practice of the Soviet Union and Iran since the conclusion of the Treaties of 1921 and 1940. In the jurisprudence of both countries at that time, the Caspian Sea appeared as a so-called closed sea. As Iran and the USSR were exclusive coastal countries, they saw the Caspian Sea as a Soviet–Iranian "closed sea." Accordingly, they took the position that the Caspian Sea was under the full sovereignty of the littoral states and remained closed for access by other countries. However, the former littoral states differed among themselves in the interpretation of the concept of a "closed sea."

The concept of the Caspian Sea being a closed sea, founded on the Russian legal doctrine concerning the Caspian Sea, was established in the nineteenth century[28] and was the principle followed throughout the Soviet period.[29] Extensive Soviet[30] and foreign literature[31] from this period represented a legal assessment of the Caspian Sea as a closed sea. An identical legal understanding of a closed sea was applied by the USSR to the Black Sea.[32] A most significant feature of the Soviet legal doctrine of the closed sea was the recognition of the exclusive sovereignty of the coastal states. According to this approach, the coastal countries are allowed to define through an international agreement the legal status and regime of the closed sea. Thus, the contracting states were allowed to mutually determine the rights and obligations regarding the use of the sea. In the case of absence of such an agreement, the states exercised their sovereignty within the territorial waters and the regime of the central parts of the water basin resembled the regime of the high sea. The Soviet closed sea concept was considered at the international level as restricted to commercial and military activities in certain maritime sectors of the representatives of the coastal states.[33] Thus, the introduction of this concept into the draft of the Geneva Convention on the High Seas was prevented by the United States, Britain, and other countries. Iran supported the closed sea doctrine, both in its national legislation and on an international level. In 1955, Article 2 of Iran's National Law on Exploration and Exploitation of the Continental Shelf of 1949 was supplemented by the

---

[28] See: (Mamedov 2001), p. 126.
[29] See: (Buttler 1971).
[30] See: (Belli 1940; Kozhevnikov 1957).
[31] See: (Nguyen 1981; Brown 1970).
[32] See: (Darby 1986).
[33] See: (Buttler 1971).

3.5 Legal Interpretation of the Soviet–Iranian Treaties

provision that international rules regarding the closed seas are applicable in the Caspian Sea.[34] Some authors claim that the only purpose of this controversial legal concept was to emphasize the difference between the regime of the Caspian Sea and those of other waters.[35] In 1974, Iran officially reaffirmed its previous assessment of the Caspian Sea being a closed sea by pointing out that the notion of "closed sea" shall not be confused with the concept of the enclosed sea, defined in UNCLOS (Article 122).[36] The difference between the concept of "enclosed sea" and "closed sea" is that the latter is not entirely closed.[37]

The Soviet–Iranian concept of a "closed sea" is a legal concept originally drafted by the former Caspian littoral states, which should not be confused with "enclosed sea" within the meaning of UNCLOS. According to the international law, "closed sea" means a sea that has no connection to the world ocean and is surrounded by two or more states. Closed seas are excluded from the provisions of the Convention of 1982 and thus remain entirely under the exclusive control of the littoral states, which may exercise their sovereignty over the entire sea or its parts without any restriction. This position was represented in intergovernmental negotiations by Kazakhstan right after the dissolution of the USSR. The basic difference of the legal positions regarding the "sea related" status of the Caspian Sea represented by Kazakhstan's and Russia–Iran's concepts was the classification proposed by Kazakhstan to define the Caspian Sea as an "enclosed sea" according to the understanding of UNCLOS. An official letter from the Permanent Representative of Kazakhstan in 1997 to the Secretary General of the United Nations[38] proposed to apply individual provisions of UNCLOS to the Caspian Sea, considering the specific characteristics of the Caspian Sea. The seabed of the Caspian Sea and its resources would be delimited by all coastal states along the middle line. Each coastal state would perform its exploration and exploitation in its economic zone independently. The exploitation of resources located within the economic zones of two different Caspian states should be the subject of a separate bilateral agreement. The parties should agree on the width of coastal waters and the fishing zone remaining under national jurisdiction. The areas seaward of these zones should remain open for the navigation of the ships of the Caspian littoral states. Also, the airspace over the Caspian Sea is open to all aircraft flying along the agreed routes. Fisheries and use of other biological resources would be carried out in fishing zones according to quota and a licensing system. Furthermore, Kazakhstan suggested that states surrounding the Caspian Sea should enjoy

---

[34]National legislation and treaties relating to the law of the sea, New York: UN, 1974. XXXIV, p. 151 Document Symbol: ST/LEG/SER.B/16 (Article 2 note).

[35]See: (Mehdiyoun 2000).

[36]For the purposes of the UNCLOS Convention, "enclosed or semi-enclosed sea" means a gulf, basin or sea surrounded by two or more States and connected to another sea or the ocean by a narrow outlet or consisting entirely or primarily of the Territorial Seas and exclusive economic zones of two or more coastal States.

[37]See: (Nordquist 1985).

[38]UN Doc. A/52/424, UN Doc. A/51/529.

the right to use Russia's waterways to get access to other lakes and oceans, upon a separate agreement with the Russian Federation.

Right before the conclusion of the Caspian Sea Convention, merely Iran, which called for an equal division of the Caspian Sea among the five littoral states, seemed to still support the concept of a "closed sea." Neither Kazakhstan nor Russia classified the Caspian Sea according to existing legal concepts pertaining to sea. With the completion of the North Caspian agreements[39] at the end of the 1990s and with the tough negotiations on the status of the Caspian Sea, the two neighboring states seemed to be satisfied with the unclear status of the Caspian Sea and remained inconsistent only in their positions on the question on the Trans-Caspian pipelines. This legal concept of the Caspian Sea being a "sea" has not been upheld in the Caspian Sea Convention as concluded in 2018.

3.5.2 Caspian Sea as a "Lake" in Legal Terms

One of the most determined advocates of the concept of the Caspian Sea as an international lake is Azerbaijan. Its legal position was presented to the other coastal states by the end of 1994 in the form of a Draft Convention.[40] The Caspian states were encouraged to practice mutual understanding in the elaboration of a new legal status for the Caspian Sea. Due to its physical-geographical conditions, the Caspian Sea was defined as a border lake, being an internal continental closed basin without a natural connection to the ocean. Azerbaijan proposed to divide the Caspian Sea into national sectors. A sector should be understood as a part of the water area and of the seabed adjacent to the coastal state seaward of the coastal waters, being an integral part of the coastal state. Thus, proposed sectors shall fall under the states' sovereignty. Their borders should be delimited based on the middle line principle, wherein each point is equidistant from the coast. According to international law, the determination of the legal regime of a border lake is left exclusively to its neighboring states because there is no international convention that would regulate this issue in a universally binding way. Border lakes are part of the internal waters of a country. Usage rights, environmental protection, water management, shipping etc. are left to bilateral or multilateral agreements between the coastal states. UNCLOS enjoys no direct application to a boarder lake; however, some of its legal principles may serve as guidance.

With only a few exceptions in the international practice, border lakes are divided among coastal states. In respective intergovernmental agreements, states define the borders of their national sectors, which fall under their sovereignty. There are some standard methods for sector delimitation: "thalweg," "coastal line," and "middle

---

[39]See Sect. 4.4.

[40]Draft of the Convention in Records of the Foreign Office of the Azerbaijan Republic, in: Mamedov (2001), p. 224.

line." Thalweg, the line of lowest elevation within a watercourse, is a method often used for the delimitation of international rivers and rarely for border lakes.[41] The principle of coastal line is usually applied in the practice of colonial countries and was later replaced by the principle of the middle line.[42] Other delimitation methods applicable to international lakes are astronomical line,[43] straight line,[44] coastal line,[45] and historical boundaries.[46] In the international practice of border lakes, the generally used method[47] is the geographical middle line[48] and in the case of complex coastlines (with islands, peninsulas, etc.) the formal middle line.[49] However, there is no uniform practice in international law to delimit border lakes using the middle line.[50]

Lake Constance (in German known as Bodensee) has a yet different legal status. There is, however, no consent from the three littoral states regarding how to define it. After the dissolution of the Holy Roman Empire of the German nation in 1806, only the treaties on the use of Lake Constance were adopted, but the boundary lines were left undetermined.[51] The unclear and conflicting declarations of intent expressed by the shore states do not allow either a determination of a condominium regime or a delimitation of Lake Constance.

The only example of a border lake, which is regulated with a condominium status, is Lake Titicaca.[52] According to the Agreement for the Boundary Correction of

---

[41]The US Supreme Court Minnesota v Wisconsin [1920] 252 US 273; Concerning Borgne See, in: The US Supreme Court Louisiana v Mississippi, 1906, 26 p. Ct. 408 ,571 and 202 US 1, 50, 58.

[42]In case of the Lake Malawi: Anglo-German Agreement of 1890 and Luso-British Agreement of 1891, Anglo-Portuguese Agreement of 1954; In case of the Caspian Sea: Treaty of Turkmenchay of 1828, Treaty of 1940.

[43]In case of the Lake Victoria (Uganda, Kenya, Tanzania): Anglo-German Agreement of 1890; partly in case of the Lake Chad (Chad, Cameroon, Niger, Nigeria): Anglo-French Conventions of 1898 & 1904 & 1906; In case of the Lake Prespa: Florence Protocol of 1926; In case of the Lake Tanganyika (Tanzania, Burundi, Congo): English–Belgian Protocol of 1924.

[44]In case of the Lake Ohrid: Florence Protocol of 1926; In case of the Lake Doyran: Border Treaty between Yugoslavia and Greece of 1959; Lake Khanka: Convention of Peking of 1860.

[45]In case of the Lake Ladoga: Moscow Peace Treaty of 1940.

[46]In case of the Neusiedler See: Treaty of Trianon of 1920.

[47]See: (Verdross and Simma 1984), §1055.

[48]In case of the Lake Malawi (Nyasa): British–Portuguese Agreement of 1954; In case of the Lake Lugano: Switzerland–Italy Agreements.

[49]In case of the Lake Geneva: Convention between Switzerland and France on the Determination of the frontier in Lake Geneva of 1953; In case of the Lake Albert (Uganda and DRC): London Agreement of 1915.

[50]See: (Pondaven 1972).

[51]See: (Schweiger 1995).

[52]See: (Barsegov 1998), p. 8; In case of the Lake Titicaca (between Peru and Bolivia): Lapas Protocol of 1925 and of 1932; Originally also in case of the Lake Mirim (Treaty between Brazil and Uruguay Modifying their Frontiers on Lake Mirim and the River Yaguaron, and Establishing General Principles of Trade and Navigation in those Regions of 1909); In case of the Lake Skadar (between Yugoslavia and Albania): Florence Protocol of 1926.

17 September 1909 between Peru and Bolivia (Tratado de Rectificación de Frontieras) and its Additional Protocol of 2 June 1925, the lake was originally divided among the shore states.[53] This regulation was, however, amended afterward by a Treaty of 19 February 1957 (Convenio Para. el studio economico preliminar de aprovechamiento de las aguas del Lago Titicaca), regulating the efficient use of waters. The treaty introduced "indivisible and exclusive condominium over the waters of Lake Titicaca" between Peru and Bolivia "without amending the fundamental conditions of navigation, fisheries and water column" (Article 1).[54]

The consideration of the Caspian Sea as an international lake can be summarized as follows: originally, Azerbaijan defined the Caspian Sea as a border lake and called for its division into national sectors along the middle line. This request *de lege ferenda* was followed by a recognition of primarily western oil companies to work in the claimed Azerbaijani sector of the Caspian Sea according to the so-called Contract of the Century. With the completion of the North Caspian treaties, Azerbaijan's position regarding the final status of the Caspian Sea was that regardless of any applicable legal concept, the use of Caspian resources should not be hampered. This approach of the Caspian Sea as a lake has not been upheld by the Caspian Sea Convention.

3.5.3 The Caspian Sea as "Condominium" in Legal Terms

According to the theory of condominium, a border sea is under the joint political authority of all coastal states, which are equally sovereign in the sea. This view was represented by Russia[55] and Iran in the early 1990s. Also, Turkmenistan occasionally supported the regime of the Condominium for the Caspian Sea,[56] but its position was often subject to change.[57] Russia and Iran claimed that the existing status of the Caspian Sea shall be defined based on the Soviet–Iranian Treaties of 1921 and 1940, which do not provide for its division. On the contrary, the diplomatic notes exchanged upon the conclusion of these agreements refer to the Caspian Sea as a "Soviet–Iranian Sea." In its note to the Secretary General of the United Nations on the legal status of the Caspian Sea, Russia emphasized the need for its common management and the use of its natural resources by all coastal countries, where no

---

[53]See: (Garcia 1996).

[54]See: (Mendoza Pastor 1958), pp. I et seq.

[55]See: (Elferink 1998a, b).

[56]UN Doc. A/53/453 from 2nd October 1998.

[57]UN Doc. A/52/93 from 17th March 1997, in the letter to the UN Secretary General they reported the joint declaration of 27 February 1997, where they mutually recognized the right of exploitation of natural resources of the Caspian Sea. UN Doc. A/55/309 from 22nd August 2000, in the letter to the UN Secretary General Turkmenistan can also accept this principle (of sectorial division of the Caspian Sea), just as it accepted the earlier concept of a "common sea," p. 6.

unilateral actions could be considered legal.[58] Furthermore, the note said that Russia would retain the right to take any appropriate and necessary measures to restore the proper regime of the Caspian Sea. Iran, in its letter to the UN Secretary General, also opposes the division of the Caspian Sea.[59] Since, as claimed by Iran, the Soviet–Iranian agreements do not provide for any boundaries in the Caspian Sea, any attempt for its division would be illegal.

The concept of condominium is controversial in international law except in some historical cases.[60] In its decision on the Gulf of Fonseca, the International Court of Justice held the condominium regime to be appropriate in case of a dispute among the successor states.[61] However, it pointed out that this principle is applicable to an area that had previously been under the sovereignty of a single state. In another decision concerning Lac Lanoux regarding the territorial dispute between Spain and France on whether the lake was a condominium, the International Court of Justice identified several basic conditions under which a lake could be described as a condominium.[62] First of all, it emphasized that there must be a "clear and convincing" consent of the contracting parties on the existence of a condominium.[63] Both ICJ decisions contain requirements that were never fulfilled in the legal practice of the Caspian states and therefore exclude their application to the Caspian dispute settlement.

---

[58] UN Doc. A/49/475.

[59] UN Doc. A/52/913.

[60] Convention of Gastein of 1865; Cromer–Ghali Agreement of 1899; Anglo-Egyptian Condominium of Sudan (1898–1955); Anglo-French Condominium of New Hebrides (1914–1980); See also: (Brownlie 2008).

[61] ICJ Rep. 1992, pp. 350 ff. pp. 598 f. Para. 400.

[62] 24 ILM (1957), pp. 101–142.

[63] ICJ stated that "to admit that jurisdiction in a certain field can no longer be exercised except on the condition of, or by way of, an agreement between two states, is to place an essential restriction on the sovereignty of a state, and such a restriction could only be admitted if there was clear and convincing evidence. International practice does reveal some special cases in which this hypothesis has become reality; thus, sometimes two States exercise conjointly jurisdiction over certain territories (joint-ownership, co-imperium, or condominium); likewise, in certain international arrangements, the representatives of States exercise conjointly a certain jurisdiction in the name of those States or in the name of organizations. However, these cases are exceptional, and international judicial decisions are slow to recognize their existence, especially when they impair the territorial sovereignty of a State. Furthermore, the Tribunal stated that "as between Spain and France, the existence of a rule requiring prior agreement for the development of the water resources of an international watercourses can therefore result only from a treaty Spanish thesis that the necessity for prior agreement would derive from all the circumstances in which the two Governments are led to reach agreement is in contradiction with the most general principles of international law." And, "as prohibited was seen a 'right of assent' a 'right of veto', which at the discretion of one state Paralyses the exercise of the territorial jurisdiction of another."

3.5.4 North Caspian Agreements

The legal theories of the Caspian Sea as a sea, lake, or condominium had confronted each other in a legal debate for many years. The explicit acceptance of one of the concepts by the negotiating parties would have inevitably led to the regulation of the Caspian Sea's status based on such classification and the accompanying body of law. If the Caspian Sea is a "sea" in legal terms, the law of the sea would be applicable.[64] If, on the other hand, the Caspian Sea is a "lake" or a "condominium" in legal terms, then customary international law concerning, respectively, border lakes[65] or condominiums would apply.[66]

The signing of the North Caspian Agreements by Azerbaijan, Kazakhstan, and Russia, and later Turkmenistan, despite the disagreements expressed by the remaining coastal states, reflected their current position in the debate regarding the status of the Caspian Sea. In parallel, Azerbaijan and Turkmenistan conducted bilateral negotiations on the status of the resource fields lying between their coasts. It promotes the effective use of the Caspian resources without dismissing ongoing multilateral negotiations by all coastal countries on the future status of the Caspian Sea. With the signing of the Caspian Sea Convention, the question of whether the Caspian is a sea, lake, or condominium in legal terms has disappeared from the legal framework for defining the Caspian Sea's status. The Caspian Sea Convention, as adopted in 2018, reflects a preference by the coastal states for a combination of divergent legal approaches to defining the legal status of the Caspian Sea, as will be discussed later in details.

3.6 Legal Confusions in State Practice Regarding the Use of Resources in the 1990s

The divergent interpretations of the existing Soviet–Iranian treaties, being the legal basis for the status of the Caspian Sea and thus for the rights and duties of the coastal states concerning the use of its natural resources, resulted in conflicting unilateral and multilateral actions by the Caspian littoral states. In September 1994, the so-called Contract of the Century between the Azerbaijan International Oil Consortium and several international oil companies was signed for the exploitation of large offshore oil fields in the Caspian Sea (Guneshli, Chirag, Azeri, Kyapaz/Serdar), creating a joint venture, AIOC (Azerbaijani International Oil Consortium). Azerbaijan's sovereignty claims regarding the use of the Caspian resources were introduced into its state constitution of 1995. Its Article 11 states that "for the

---

[64]See: (Bodenbach 2008).
[65]See: (Romano 2000), pp. 145–161.
[66]Ibidem.

3.6 Legal Confusions in State Practice Regarding the Use of Resources in the 1990s

territory of Azerbaijan entails Internal Waters and Azerbaijani sector of the Caspian Sea."

The conclusion of the Contract of the Century emphasized the lack of clarity of the existing legal status of the Caspian Sea and turned into a burning challenge to the relations between the Caspian states. The unilateral action of Azerbaijan raised sharp criticism from other Caspian littoral states on one hand, but at the same time it opened the way for similar actions of other coastal states. In reaction to the Contract of the Century, the uninvolved coastal states expressed the opinion that any unilateral actions regarding the Caspian states might be legitimate only in case of a joint decision by all coastal countries.[67] At the same time, they also reached for actions to use Caspian resources that were uncoordinated with other coastal states. Turkmenistan, already in 1993, enacted a law providing for the establishment of Internal Waters, a 12-nautical-mile wide territorial sea and an exclusive economic zone. The new regulation covered also the Caspian Sea, which clearly referred to Turkmenistan's sovereignty claims in this area. Kazakhstan and Azerbaijan as well as Kazakhstan and Turkmenistan issued joint declarations mutually awarding each other the right to exploit the natural resources of the Caspian Sea.[68] Kazakhstan and Azerbaijan expressed sovereignty claims over the Caspian Sea also in a multilateral initiative, the so-called Ankara Declaration of 29 October 1998, regarding the exploitation of Caspian resources and their transportation.[69] The document signed by Azerbaijan, Georgia, Kazakhstan, and Uzbekistan emphasized the great importance of the Caspian resources and their transportation for these countries. Special support was expressed for the building of the Trans-Caspian gas and oil pipelines.

Russia's originally critical position toward any unilateral use of Caspian resources by coastal states changed with the adoption of the North Caspian Agreements on the use of Caspian resources, which were concluded despite opposition from Iran and Turkmenistan. Russia's change of mind about the legitimacy of the allocation of rights for the use of the natural resources in the Caspian Sea has been justified by the fact that previous negotiations were ineffective and too long. Russia preferred to continue the multilateral efforts to settle Caspian's status[70] but proposed that all coastal states simultaneously carry out separate negotiations regarding each legal regime, like navigation, use of resources, environmental protection, etc., in the Caspian Sea.

---

[67]UN Doc. A/49/475, annex, October 5, 1994 (Russian Federation), UN Doc. A/51/59, annex, January 27, 1996 (joint statement of Iran and Russia October 30, 1995), UN Docs. A/51/73, annex, March 1, 1996 (joint statement by Russia and Turkmenistan August 12, 1995); A/51/138, Annex II, May 17, 1996 (joint statement of Kazakhstan and Russia April 27, 1996), UN Doc. A/52/324, Annex, 8 September 1997 (Iran).

[68]UN Doc. A/51/529 of 21 October 1996 (Azerbaijan–Kazakhstan); UN Doc. A/52/93 of 17 March 1997; UN Doc. A/52/324 of 8. September 1997 (Kazakhstan–Turkmenistan).

[69]UN Doc. A/C.2/53/9 of 3rd December 1998.

[70]See: (Kolodkin 2002b).

Originally, Iran held a clearly negative position toward unilateral actions in the Caspian Sea.[71] A very serious situation, with the involvement of military threats, arose around the Araz–Alov–Sharg oil field (known in Iran as Alborz). On August 18, 1998, Azerbaijan officially announced that it had plans to carry out activities in this field, to be conducted in collaboration with some oil companies. Iran first pointed out to Azerbaijan that it would need Iran's consent to lawfully carry out the planned activities.[72] A few days later, on July 23, 2001, a British Petroleum (BP)-operated research vessel exploring this contested offshore field suspended its exploration activities under the pressure of Iranian military vessels and aircraft.

Despite its originally expressed criticism, Iran itself consented to unilateral actions in the Caspian Sea. In December 1998, the signing of a contract for geological and geophysical exploration between Iran and Shell and Lasmo oil companies was announced. The respective area was regarded by Azerbaijan as part of its own sector on the Caspian Sea; thus, it sent a protest letter to the UN Secretary General.[73] On May 24, 2000, Iran undertook another unilateral action by issuing a national legal act authorizing the National Iranian Oil Company (NIOC) to explore and exploit oil and gas resources in the Caspian Sea.[74] The NIOC was permitted to conclude contracts with both local and foreign companies.

In the 1990s, the Caspian littoral states represented alternately contradictory positions and performed clashing actions. On the one hand, they sharply criticized the neighboring countries, whose unilateral actions affected areas that they regarded as belonging to their national sectors of the Caspian Sea. At the same time, at the turn of the century at the latest, they all entered into an agreement with regional and external partners regarding the use of resources within sectors of the Caspian Sea claimed by them as national. Simultaneously, at the level of political statements, the littoral states agreed to look toward a legal compromise and a possible extensive cooperation. The expression of the political will to cooperate was a continuation of the interstate negotiations on the Convention on the legal status of the Caspian Sea, which have been carried out continuously since the mid-1990s until today. The multilateral approach was also expressed when aiming for the conclusion of agreements on separate legal regimes of intergovernmental cooperation in the Caspian Sea. This concept was enforced with the conclusion of the Tehran Convention of 2003 and the Caspian Security Agreement of 2010. Therein, where the compromise among all littoral states appeared to be temporarily unreachable—as it was the case regarding the use of the natural resources of the northern Caspian Sea—states undertook legal actions bilaterally regulating their relations in respective areas of

---

[71] Iran's rejecting reaction to the declaration of Kazakhstan and Turkmenistan (UN Doc. A/52/324 of 8th September 1997), to the announcement of extraction and exploitation of oil from the Cheragh reservoir by Azerbaijan (UN Doc. A/52/588 of 25th November 1997), "Ankara Declaration" condemning the trans-Caspian pipelines proposal (UN Doc. A/54/788 from 9th March 2000.).
[72] UN Doc. A/56/304 of 17th August 2001.
[73] UN Doc. A/53/741 of 14th December 1998.
[74] Official Gazette no. 16114, of 25th April 1379 of 25th June 2000. In: (Ranjbar 2004), p. 88.

the Caspian Sea, despite opposition from the remaining coastal states. At the same time, as expressed directly by all treaties related to territorial sectors or regimes of usage, none of them should prevent a future comprehensive agreement on the legal status of the Caspian Sea, but they should rather be regarded as part of the final overall mutual agreement of all the five coastal states. The Caspian Sea Convention introduced a new understanding of the legal status of the Caspian Sea.

References

Antonowicz L (1991–1992) The disintegration of the USSR from the point of view of international law. Polish Yearb Int Law 19
Barsegov J (1998) Kaspij v mezhdunarodnom prave i mirovoi politike (Caspian Sea in the international law and global policy). Institute of International Politics and International Relations, Russian Academy of Sciences, Moscow
Belli V (1940) Navy international-law manual book, 2nd edn. Leningrad, Moscow
Bodenbach E (2008) Die völkerrechtliche Einordnung internationaler Seen unter besonderer Berücksichtigung des Kaspischen Meeres. Peter Lang Verlag, Frankfurt am Main
Britannica E (2005) Encyclopedia Britannica Online. [В Интернете] Available at: http://www.britannica.com/EBchecked/topic/452741/Persia [Дата обращения: 1 July 2020]
Brown L (1970) Public international law. Sweet & Maxwell, London
Brownlie Q (2008) Principles of public international law, 7th edn. Oxford University Press, Oxford
Buttler W (1971) The Soviet Union and the law of the sea. John Hopkins Press, Baltimore
Darby J (1986) The Soviet doctrine of the closed sea. San Diego Law Rev 23:685
Elferink AO (1998a) Denmark/Iceland/Norway-bilateral agreements on the delimitation of the continental shelf and Fishery Zones, 13th edn. IJMCL
Elferink AO (1998b) The legal regime of the Caspian Sea. Are the Russian arguments valid?. In: Wolfrum R (ed) The legal foundation of the New Russia. Norwegian Institute of International Affairs, Oslo
Garcia W (1996) Limites de Bolivia, 2nd edn. Libreria editorial "Juventud", La Paz, Bolivia
Gromyko A, Kovalev P, Sevostyanov S, Tihvinskiy S (1985) Diplomatic dictionary. Nauka, Moscow
Gull K (1960) From the historical geographical researches of Caspian Sea. Izvestija Akademii Nauk Azerbeidschanskoj SSR (News of the Academy of Sciences of Azerbaijan SSR), Issue Series of Geog.-Geology Sciences 2
Ipsen K (2004) Völkerrecht, 5th edn. CH Beck, München
Jiloe P (1960) About the appellations of Caspian Sea.. Izvestuja Akademii Nauk Azerbeidschanskoj SSR (News of the Academy of Sciences of Azerbaijan SSR), Issue Series of Geog.-Geology Sciences 42, Baku
Kolodkin A (2002a) On the legal regime of the Caspian Sea. Oil Gas Law 2(44)
Kolodkin A (2002b) Ne zhdat' u morya pogody (Don't whistle for a wind). Izvestya
Kozhevnikov F (1957) Mezhdunarodnoye pravo (International law). MGIMO University, Moscow
Law A. S. o. I. (1978) International legal materials. International Legal Materials, Issue 17
Law A. S. o. I. (1992) International legal materials. International Legal Materials 1:147–154
Mamedov R (2001) International legal satus of the Caspian Sea; issues of theory and practice. Turkish Yearb Int Relat 32:217–259
Mehdiyoun K (2000) International law and the dispute over ownership of oil and gas resources in the Caspian Sea. Am J Int Law 94:179–189

Mendoza Pastor C (1958) El Aprovechamiento de las agues del lago Titicaca y los problemos juridicos que plantea (The exploitation of the water of Lake Titicaca and legal problems). Pontificia Universidad Católica del Perú, Lima, Peru

Nguyen NM (1981) International maritime law. Moscow (s.n.)

Nordquist M (1985) United Nations Convention on the law of the Sea 1982 – a commentary. In: Center for Oceans Law and Policy. Martinus Nijhoff, Dordrecht

Polat N (2002) Boundary issues in Central Asia, 28th edn. Transnational, Ardsley

Pondaven C (1972) Les Lacs-frontiere. Pedone, Paris

Ranjbar R (2004) Das Rechtsregime des Kaspischen Meeres und die Praxis der Anrainerstaaten. Nomos, Baden-Baden

Romano C (2000) The Caspian Sea and international law: like oil and water. In: Mirovitskaya N, Ascher W (eds) The Caspian Sea: a quest for environmental security. Kluwer, Dordrecht

Schweiger K (1995) Staatsgrenzen im Bodensee und IGH-Statut in Bayerische Verwaltungsblätter. Zeitschrift für öffentliches Recht und öffentliche Verwaltung, 3

Schweisfurth T (1992) Vom Einheitsstaat (UdSSR) zum Staatenbund (GUS). Juristische Stationen eines Staatszerfalls und einer Staatenbundentstehung. HIJL 52:541–702

Schweisfurth T (1996) Das Recht der Staatensukzession. Die Staatenpraxis der Nachfolge in völkerrechtliche Verträge, Staatsvermögen, Staatsschulen und Archive in den Teilungsfällen Sowjetunion, Tschechoslowakei und Jugoslawien. In: Fastenrath U, Schweisfurth T, Ebenroth C (eds) Das Recht der Staatensukzession. C.F. Müller Verlag, Heidelberg

Verdross A, Simma B (1984) Universelles Völkerrecht, 3rd edn. Duncker & Humblot, Berlin

Vylegjanin A (2000) Basic legal issues of the management of natural resources of the Caspian Sea. In: Mirovitskaya N, Ascher W (eds) The Caspian sea: a quest for environmental security. Kluwer, Dordrecht

Chapter 4
Cooperation Levels in Caspian States Practice in the 1990s Until 2018

4.1 Challenges for the Caspian Region After the Dissolution of the Soviet Union

The incompleteness of the Soviet–Iranian agreements and their conflicting interpretation raised doubt about the legality of the activities of Caspian littoral states in the 1990s. Actions done by coastal states, often taken in consideration of the mutually exclusive political and economic national interests of individual states, have an effect on all the strategic areas of cooperation in the Caspian Sea, including the questions of setting of state borders, operating of ships and fisheries, and degradation of natural resources and their transportation. This precarious situation alters already existing legal problems and creates new ones. The establishment of a uniform legally binding document regulating the collaboration of states in the Caspian Sea basin was of tremendous importance for the political security of the entire region and its future economic development. The stability of and clarity on the legal situation in the Caspian region is no less important for securing international investments in the extraction of Caspian resources, obtaining foreign loans, as well as the purchase of shares or exploitation rights to Caspian oil and gas fields. The introduction of legal standards and rules oriented toward peace and international legal standards would make an important contribution toward the prevention of military solutions.

Among the large number of existing questions related to international law in the post-Soviet space, the one concerning uncertainty about the existence and development of the state maritime borders of the five coastal states around the Caspian Sea was especially important. No clear statement on this issue was made in the treaties between Iran and the Soviet Union in the years 1921, 1935, and 1940. One could not explicitly determine whether the arrangements made by the coastal states in the past can be regarded as a delimitation of the basin or whether the Caspian Sea has been declared by the concerned states as an area that should be commonly used by all the states. This question was of fundamental importance for the successful solution of other problems concerning the Caspian Sea. Borders define the area of state sovereignty and thus narrow the scope of state jurisdiction. If the Caspian Sea became divided, issues such as the route of trans-Caspian pipelines, measures to protect the basin's environment, transportation, etc. would depend solely on the individual and independent decisions of the individual coastal states. If, however, the Caspian Sea was subjected to a joint administration regime, it would be in common use, and all decisions and actions must be agreed upon with the remaining coastal states. The existing unclear status contributed to the legal uncertainty over the status of the Caspian Sea and created increasing instability in the region.

Sea transport is one of the most important means of developing intergovernmental economic ties in the Caspian Sea. In international law, there is a general, broadly recognized principle of freedom of navigation on the seas for each state.[1] The principle of transit, freedom to enter a port facility, and freedom to transport goods are also valid for navigation on international rivers and lakes.[2] Such principles are indeed part of the legal system. However, direct legal obligations derived from them can only be formed within narrow limits. To be directly applicable, they need to be defined in other legal norms. Already, the Soviet–Iranian treaties guaranteed the freedom of navigation on the whole Caspian Sea, however with the crucial limitation that this applies only to vessels flying the flag of one of the Caspian states. For the current trade needs, the question of the international legal order of commercial and naval shipping on the Caspian Sea remains hardly regulated. Therefore, there was an urgent need for an intergovernmental regulation of commercial shipping on the Caspian Sea that would take as a basis currently applicable norms of international law.

The Caspian Sea has abundant fish stocks. The world famous Caspian sturgeon has become the basis of an independent branch of industry. In recent years, the abundance of fish strongly decreased because of pollution and overfishing. Even the existence of the sturgeon is endangered. A comprehensive protection regulation is urgently needed. Under the auspices of the Commission on Aquatic Bioresources of the Caspian Sea, a draft Agreement on Conservation of aquatic bioresources of the Caspian Sea and their management has been discussed since 2003. Simultaneously, the coastal states discussed the possible contents of a Protocol on Conservation of Biological Diversity ancillary to the Tehran Convention, which was adopted and signed at the Fifth Meeting of the Conference of the Parties in Ashgabat, Turkmenistan, on May 30, 2014 but still did not enter into force.

The existing and forthcoming large-scale oil and gas drilling threatens to deal with another heavy blow to the Caspian's ecosystem. The proposal to construct another oil pipeline on the floor of the Caspian Sea has not only been politically highly controversial but could also expose the Caspian ecosystem to another significant danger. The lack of clear intergovernmental provisions in this regard required urgent regulation.

---

[1] Article 90 UNCLOS.
[2] Article XIV Helsinki Rules (1966).

Increased cooperation among the Caspian littoral countries has been an essential prerequisite for the successful implementation of all normative regulations of the Caspian states, with ten million inhabitants living in direct dependence on the Caspian Sea. To the same extent, the necessity of close cooperation applied to the use of the living and nonliving natural resources of the Caspian Sea, the protection of the fragile Caspian environment, and any action concerning the economic development of the region. A normative reform could have completely changed the political situation in the Caspian region, and close cooperation in all subject areas would have been necessary for a successful reform on the international, national, regional, and local levels.

Intergovernmental cooperation in the area of lawmaking seemed to be even more urgent as the existing problems become even more explosive because of the emergence of previously unknown threats. Drugs and illicit arms trafficking, illegal immigration, fish poaching, and organized crime, classified as crimes under international law, have become so widely spread on the Caspian Sea that the community of the Caspian states has been greatly challenged. Not only the security and sea traffic but also the environment of the Caspian Sea was threatened by maritime terrorism and piracy. The new urgently needed regulations should be made to prevent these crimes in the Caspian Sea.

Some of the abovementioned issues requiring a new regulation arose because of the absence of rules or their inconsistent interpretation. Furthermore, they lead to inconsistent, even contradictory and mutually exclusive, actions of neighboring states, and thus they jeopardize the situation in the entire region. Being aware of the danger since the 1990s, the littoral states continued their search for the ways to a regional compromise. It was achieved step by step at the bilateral and multilateral levels targeted at both the complex approach and the regulation of the individual "subject areas" and was finally settled by the adoption in 2018 of the Convention on the Caspian Sea Status.

4.2 Peaceful Settlement in International Law

The interstate conflict that had been observed in the Caspian Sea since the 1990s is subject to international legal regulations established for the peaceful settlement of international disputes.[3] The international dispute settlement procedures do not provide a uniform definition of the term.[4] However, it is possible to define some aspects of an international dispute as follows:

---

[3]On the concept of "international disputes" see: (Caron and Shinkaretskaya 1995), p. 309.
[4]Art. 36(2) IJC Statute; Arts. 2003 et seq. of NAFTA Agreement.

An international dispute can take the following forms: disagreement over a fact or a legal or political issue, conflict between a number of parties or situation when a demand or a claim of a party is denied or disputed by the other party or the other party submits a counterclaim.[5]

An international legal dispute is a disorder of intergovernmental relations. A distinction between legal and political aspects is very difficult but extremely important. Depending on what aspects prevail, different settlement procedures are performed. For example, Article 65 of the International Court of Justice (ICJ) Statute provides that the ICJ can give its so-called advisory opinions only to a legal question. Therefore, the international legal doctrine tries to adequately separate disputes over political issues from the vital interests or honor of a state. Disputes over legal questions are related to the interpretation or application of law. Whether a dispute must be regarded as justifiable depends more on the willingness of states to subject themselves to court jurisdiction and less on the possible inadmissibility of dispute settlement proceedings before a court. In principle, any dispute can be decided under the rules of international law. However, not all disputes arising from these decisions are settled. For example, a supposed claim can be dismissed because of the lack of an international legal basis for a claim, and the dispute remains unresolved despite the fact that a decision was taken.[6] A hint about the legal validity of the customary law of obligation to peacefully resolve disputes is provided by successive resolutions of the United Nations (UN), such as the Friendly Relations Declaration,[7] the Manila Declaration,[8] Resolution 40/9, the Declaration on the Prevention and Removal of Disputes and Situations Which May Threaten International Peace and Security and the Role of the United Nations in This Field,[9] and the United Nations Decade of International Law.[10] The norms of general international law with respect to dispute resolution are also enshrined in the Charter of the United Nations:[11]

> All Members shall settle their international disputes by peaceful means in such a manner that international peace and security, and justice, are not endangered (Article 2(3)).

These positive obligations of states to peacefully settle disputes imply that contending parties have the choice of the means to settle a dispute. They can make use of the UN Charter provisions, which provide for a diplomatic process as well as international arbitration and jurisdiction, and define other procedures. The individual processes are not ordered hierarchically, but their application in practice depends on the circumstances of every dispute. The parties to any dispute that are likely to endanger international peace and security first seek a solution through negotiation,

---

[5]See: (Land 2000), p. 19.
[6]See: (Kunz 1968), p. 684.
[7]Annex to GA Res. 2625 (XXV), 1970.
[8]Annex to GA Res. 37/10, 1982.
[9]GA Res. 41/92, 1986.
[10]GA 44/23, 1989.
[11]See: (Kimminich 1997), p. 282.

inquiry, mediation, conciliation, arbitration, judicial settlement, resort to regional institutions or arrangements, or other peaceful means of their own choice.[12]

4.3 Five-Party Negotiations on the Convention on the Legal Status of the Caspian Sea from 1990 Till 2018

In the completely new geopolitical situation in the Caspian region after the dissolution of the Soviet Union, there was an urgent need for a new legal regulation in terms of both status and regime and, thus, also of the rights and obligations of the states parties regarding the use of the Caspian Sea, including their waters above the seabed, the subsoil, the natural resources, and the airspace above the sea. The adoption of a five-party document regulating the legal status of the Caspian Sea in a binding manner would be the basis for a free undertaking of future legal obligations by the coastal states regarding associated issues. Legal clarity and, consequently, stability are also of enormous importance for ensuring international financial investment in the extraction of Caspian resources, foreign loans, and the purchase of shares or exploitation rights to Caspian oil and gas fields. The introduction of a peace-oriented agreement respecting international legal standards would have an important contribution to preventing future military interventions.

The question of a comprehensive regulation of all aspects of the status and regime of the Caspian Sea was raised for the first time by the delegation of Azerbaijan during the intergovernmental conference to resolve emerging problems in the Caspian region (Tehran, September/October 1992).[13] The first stage of the negotiations, which started between the Caspian countries during this conference, was characterized by the multilateral and equal participation of all the neighboring countries. The draft agreement to establish a cooperation organization among the Caspian states, prepared by Iran and presented during that conference, met with little interest from the other countries. Instead, they agreed on certain areas of cooperation, such as protection and sustainable use of natural resources and establishment of sea routes respecting the interests of all the states.[14] As a result of the conference, the so-called Committee of Biological Resources, consisting of envoys from all the neighboring states, started its work.

Cooperation at the level of the foreign ministries of all the five littoral states aiming to conclude an agreement covering all issues concerning legal status and regime was secured during the Almaty Conference in May 1995. At that time, the so-called Working Group, composed of the deputy ministers of foreign affairs of all the five littoral states, was established as a mechanism for continuous negotiations

---

[12]UN Charter, Article 33.

[13]See: (Mamedov 2001), pp. 217–259.

[14]Joint Communiqué of the Caspian States Representatives of 4.12.1992, in: Records of the Foreign Office of the Azerbaijan Republic.

concerning the legal status of the Caspian Sea.[15] At its meeting, the Working Group developed a draft of the Caspian Status Convention. This should have provided for a legal basis for other multilateral agreements among the coastal states to regulate various issues relating to the Caspian Sea. The principle of consensus was announced as an exclusive method for the approval of all future agreements regarding the Caspian Sea, a result of the foreign ministers' meeting in Ashgabat on November 12, 1996.

The most important landmarks in the negotiation process concerning the status of the Caspian Sea were the five summits participated in by the Caspian Sea heads of states.[16] During the First Caspian Summit of April 23–24, 2002, the states were not able to define the legal status of the Caspian Sea; however, they agreed not to use force to solve legal challenges.[17] During the Second Caspian Summit of October 16, 2007, in a joint declaration, Caspian states confirmed their sovereign rights over the Caspian resources, as well as agreed to limit military presence only for military vessels of the coastal states.[18] During the Third Caspian Summit on November 18, 2010, the Caspian states concluded the Agreement on Security Cooperation in the Caspian Sea.[19] At the Fourth Caspian Summit on September 29, 2014, the states adopted the Agreement on Conservation and Rational Use of the Aquatic Biological Resources of the Caspian Sea, the Agreement on Cooperation in Emergency Prevention and Response in the Caspian Sea, as well as the Agreement on Cooperation in the Field of Hydrometeorology in the Caspian Sea.[20] At the latest Fifth Summit on August 12, 2018, the Caspian states, after more than 20 years of intensive negotiations over the legal status of the Caspian Sea, were able to finally adopt a Convention on the Legal Status of the Caspian Sea.[21] The uniqueness of the negotiation process and its importance for the future of the Caspian Sea region cannot hide the fact that there were still many fundamental issues that the parties could hardly reach an agreement on.

The provisions of the entire draft of the Caspian Status Convention that had been discussed between the Caspian sea states since the 1990s aimed at explaining two different legal categories: the legal status and regime of the Caspian Sea. The Draft Caspian Status Convention proposed the following concept about the legal status of the Caspian Sea, defining the scope of authority of the individual states in the matter of water area: "States Parties shall carry out their [sovereignty] and the sovereign rights in the Caspian Sea" (Article 2 I). The recognition of states' sovereignty means acceptance of a coastal states' independent authority over a geographic area, in this case a particular zone of the Caspian Sea, which can be found in the power to rule

---

[15]See: (Mamedov 2001), p. 232.
[16]Ibp (2018), pp. 193–194.
[17]RadioFreeEurope (2002).
[18]RadioFreeEurope (2007).
[19]RadioFreeEurope (2010).
[20]RadioFreeEurope (2014).
[21]RadioFreeEurope (2018).

and make binding laws without the right of interference by other countries. Russia did not generally approve of this wording and proposed to remove the word "sovereignty." It rejected any division of the waters that would illustrate the sovereignty of coastal states over offshore waters. The Draft Caspian Status Convention came up with the term "legal regime," as opposed to legal status, and defined it as follows: "The law regime determines und rules the rights and obligations of state parties regarding the use of the Caspian Sea, including its waters above the seabed, the seabed, the subsoil, natural resources and airspace above the sea" (Article 2 II). Furthermore, "Parties use the Caspian Sea for the purpose of navigation, fisheries, use and protection of biological resources, the exploration and exploitation of the resources of the seabed and its subsoil, and for other purposes in accordance with this Convention, with the individual agreements to be settled among the Parties and with the national legislation of the States Parties" (Article 4).

The Draft Caspian Status Convention was prepared, according to its Preamble, "starting from the fact that the Caspian Sea is of vital importance to the parties." The Draft Caspian Status Convention included the principle of respect for the sovereignty, territorial integrity, political independence, and sovereign equality of all the states and the prohibition of any threat or use of force. The following principle applied to the use of the Caspian Sea for peaceful purposes; its transformation into a zone of peace, good neighborliness, friendship, and cooperation; and the settlement of all problems related to the Caspian Sea by peaceful means. Accordingly, it has been proposed—with Russia against—to introduce the principle of demilitarization of the Caspian Sea or the classification of the Caspian Sea as a demilitarized zone, which should be reserved exclusively for peaceful purposes. Furthermore, the draft included principles that determine the use regime.[22]

The provisions of the United Nations Convention on the Law of the Sea (UNCLOS), as a basic rule in maritime law, could not be applied to the Draft Caspian Status Convention as a point of legal reference because of all the Caspian states, only Russia was a party to UNCLOS and there were no indications that other states will join the agreement. Nevertheless, it was apparent that some elements of UNCLOS were carried over into the Draft Caspian Status Convention[23] since

---

[22]The principle of the prohibition of warships of non-Caspian states on the Caspian Sea, the principle of freedom and the warranty of merchant shipping safety for ships flying the flag of one of the contracting parties; the principle of denial of the right of passage to or within the Caspian Sea for ships flying the flag of a state other than a Contracting State; the principle of implementation of agreed standards and rules taken over the UNCLOS provision on the related to reproduction and regulation of the exploitation of bioresources; the principle of environmental protection of the Caspian Sea, preservation, restoration and sustainable use of its biological resources; the principle of responsibility of the states parties for having made harm to the ecological system of the Caspian Sea by causing pollution to its environment.

[23]The Draft Caspian Status Convention identified identical water categories as UNCLOS (see Sect. 5.5), does however partially define them in a different manner and therefore does not describe the water categories using an identical wording. Furthermore, with the exception of Iran, all the coastal states agreed that the seabed and its subsoil should be divided for the extraction of natural resources as well as other legitimate commercial and economic activities. This was similar to the UNCLOS

UNCLOS has been recognized as a certain embodiment of the customary[24] law of the sea.[25] Some authors even speak of a possible universal validity of UNCLOS in the near future.[26] The customary legal validity of regulations contained in some international conventions, also in relation to states that are not party to those agreements, has been repeatedly confirmed in the judicature of the International Court of Justice,[27] as well as in legal literature.[28] The Draft Caspian Status Convention declared, in its Preamble, to be bound by international law standards.

The negotiations between the coastal states were carried out from the mid-1990s till 2018. In the meantime, low confidence in the success of these negotiations made some coastal states seek for quick bilateral solutions to the exploitation of the natural resources of the northern part of the Caspian Sea already at the end of the 1990s (the so-called North Caspian treaties, which will be discussed further). However, one could also argue that it was precisely because of the conclusion of the bilateral agreements that there was no longer any urgency to search for a comprehensive legal solution to the future status and regime of the Caspian Sea and that the multilateral conclusion process was paralyzed. The negotiations on the status and regime of the

---

provisions concerning the exclusive economic zone. In addition, in accordance with the provisions of UNCLOS, all Caspian littoral States except Russia agreed that the sovereignty over the Territorial Sea would be applied in accordance with the Draft Caspian Status Convention and the rules of international law. Also, the concept of the breath of the Territorial Sea in the draft corresponds to the regulation in the provisions of UNCLOS. As one of the basic standards for their action in the Caspian Sea, the littoral states took over the UNCLOS provision on the high seas that the Caspian Sea is reserved for peaceful purposes. Any issue relating to the Caspian Sea should have been resolved by peaceful means, which is a principle enshrined in the UNCLOS. The provisions of the Draft Caspian Status Convention regarding fishing in the Caspian Sea recognized the unlimited rights of the coastal states outside the exclusive fishing zones or outside the areas of national jurisdiction. These provisions built on the corresponding provision included in the UNCLOS regarding the high seas. As for shipping in the Caspian Sea, the Draft Caspian Status Convention contained an explicit reference to internationally binding sea law provisions, of which the most significant come from the UNCLOS. Azerbaijan, Kazakhstan and Turkmenistan agree that the coastal states in which mining sites are located on the seabed and the pipeline routes are allowed to be laid down, have the right to lay trans-Caspian pipelines. To clarify the pipeline route, agreements concluded among these states are applicable. The proposal corresponded to some extent to the provisions of UNCLOS related to the rights and obligations of states related to the laying of submarine cables and pipelines.

[24] The factors of time and certain behaviors are of crucial significance for the creation of international customary law. With the current developments of international law, customary law can arise very quickly. However, two conditions are necessary: as an objective element a general practice of states, to which the subjective element occurs as the *opinion iuris sive necessitatis*. See Bernhardt (1984), p. 215.

[25] See: (Gornig and Despeux 2002), p. 6.

[26] See: (Ndiaye and Wolfrum 2007).

[27] North Sea Continental Shelf Judgement (1969), p. 41: "There is no doubt that this process is a perfectly possible one and does from time to time occur: it constitutes one of the recognized methods by which new rules of customary international law may be formed."

[28] See: (Sohn 1950), p. 1008.

Caspian Sea were continued until the signing of the Caspian Sea Convention in 2018.

4.4 Step-by-Step Conclusion of Agreements on the Use of Natural Resources

In the period from 1998 to 2004, three bilateral contracts and one trilateral agreement (so-called North Caspian Agreements) with additional protocols were concluded between Russia, Kazakhstan, and Azerbaijan to regulate the delimitation between the relevant sectors as well as the regime of exploitation of natural resources in the northern part of the Caspian Sea between these countries.

On July 6, 1998, Russia and Kazakhstan signed the first agreement regarding the division of the seabed of the relevant sectors of the Caspian Sea (Agreement Between the Russian Federation and the Republic of Kazakhstan on the Delimitation of the Seabed of the Northern Part of the Caspian Sea for the Purposes of Exercising Their Sovereign Rights to the Exploitation of Its Subsoil, further referred to as Agreement between Kazakhstan and Russia 1998). As a method of delimitation, a modified median line was used. On May 13, 2002, Kazakhstan and Russia concluded an additional protocol to this treaty, which provided for the exact coordinates of the course of the delimitation of their sectors in the Caspian Sea and contained general exploitation provisions on three oil fields. According to this additional protocol, the water column remained in common use by both parties. All issues of freedom of navigation, overflight, laying and use of cables, pipelines, etc. should be regulated under separate bilateral or multilateral agreements, to be concluded upon reaching by the states of a conclusion regarding the legal status of the Caspian Sea.

On November 29, 2001, Azerbaijan and Kazakhstan signed a delimitation agreement (further referred to as Agreement between Kazakhstan and Azerbaijan 2001). On February 27, 2003, both states concluded a supplementary protocol that delimited the seabed between the two countries along the median line. This treaty left the water column of the Caspian Sea disregarded.

On September 23, 2002, the last sector in the North Caspian region was delimited by the signing of a treaty between Azerbaijan and Russia on the delimitation of adjacent areas on the Caspian seabed (further referred to as Agreement between Azerbaijan and Russia 2002). It provides for the sovereign rights of the contracting parties over the nonliving resources in the Caspian Sea and to perform other legitimate economic activities for the exploration and resource management of the sectors of its seabed and subsoil.

The last of the North Caspian treaties was an agreement signed between all the three parties—Azerbaijan, Kazakhstan and Russia—on the convergence point of the delimitation lines of the adjacent areas of the Caspian seabed on May 14, 2003 (further referred to as Tri-Point-Border Agreement 2003).

The legitimacy of all North Caspian treaties concluded between Russia, Kazakhstan, and Azerbaijan was rejected by Iran[29] on account of their contradiction to the existing Soviet–Iranian Treaties of 1921 and 1940. Azerbaijan and Kazakhstan rejected Iran's allegations.[30] According to a generally accepted international law principle, which was codified in the Vienna Convention on the Law of Treaties of 1969, a treaty does not create either obligations or rights for a third state without its consent (Article 34). Thus, these treaties are not binding upon either Iran or Turkmenistan. However, they remain binding on their states parties. An attempt to guarantee the recognition of the legality of the North Caspian Agreements was made in the context of multilateral negotiations on the future status of the Caspian Sea. A specific regulation proposed in the Draft Caspian Status Convention providing for the principle of territorial delimitation in the Caspian Sea, upon the agreement of all Caspian states except Iran, prescribed the following: *In case where parties have already signed a relevant agreement concerning the delimitation of the seabed and its subsoil, all questions relating to the delimitation [is] to be decided in* accordance with such agreements (Article 8(9)2). This provision of the Draft Caspian Status Convention, reflecting the negotiation positions of the Caspian Sea states, included an indirect reference proposed by the North Caspian states to their bilateral agreements, which divided the nonliving resources of the northern part of the Caspian Sea between the contracting states without considering the view of the remaining Caspian coastal states. This was an attempt to secure an overall recognition of the legality of these agreements, but it was rejected by Iran.

The Turkmenistan position on the bilateral approach to the regulation of the use of the Northern Caspian Sea evolved over time. On December 2, 2014, the Agreement between the Republic of Kazakhstan and Turkmenistan on the delimitation of the bottom of the Caspian Sea between the Republic of Kazakhstan and Turkmenistan was signed and entered into force on July 31, 2015. This Agreement provides for the sovereign rights of Kazakhstan and Turkmenistan over the established sectors of the seabed and subsoil of the Caspian Sea for the exploration, development, and exploitation of resources; laying of pipelines and cables; building of artificial islands, berms, dams, overpasses, platforms, and other engineering constructions; and carrying out of other legitimate economic activities.

A few years before the bilateral approach was contested by Iran during the drafting of the North Caspian Agreements, another bilateral agreement was stipulated between Iran and Russia. On July 2, 1997, both states concluded an Agreement on Interstates Cooperation on Fisheries. This Agreement defined the following areas of cooperation: life resource breading and exploitation, management of the Caspian Sea environmental issues expending cooperation for fish products, trading in the global market evaluating the impact of traditional fishing and on the coastal fishery communities, implementing common scientific research for optimal preservation

---

[29]UN Doc. A/52/913 of 21. Mai 1998, Attachment; UN Doc. A/56/850 of 1 March 2002; UN Doc. A/56/1017 of 31 July 2002.
[30]UN Doc. A/56/927 of 18 April 2002.

and breading in the Caspian Sea life resources. For the implementation of this Agreement, a Russian–Iranian commission for fishery was foreseen. This Agreement was concluded for five years and could be automatically renewed for the same period, if none of the parties declared its termination. The Agreement was to help both countries to coordinate their fishery policies.

In the case of absence of a multilateral consent to the legal status and regime on the use of the Caspian Sea, the strategy of concluding bilateral agreements on separate issues of the legal regimes in the Caspian Sea seemed to be the only way to secure a lawful use of the Caspian resources. This solution was, however, not without controversy, especially from Iran, in regard to the regulation of the Caspian soil and subsoil resources. The drive to use the natural resources of the Caspian Sea, which provides for the bulk of states' incomes, especially of the newly independent Caspian states, could not be hindered by anyone. The coastal states were, however, still looking for securing their legality, even if only halfway, through the negotiations and, in the first phase, conclusion of separate bilateral delimitation agreements and the final adoption of the overall Convention on the Caspian Sea Status in 2018.

4.5 Step-by-Step Multilateral Regulations of the Legal Regimes in the Caspian Sea

The strategy of providing mutually agreed solutions to existing ambiguities in legal regime issues in the Caspian region required a gradual approach. For the solution to this extremely delicate and urgent problem, it was necessary to adopt one document based on consensus and adopted jointly by all the Caspian states. The first one to be adopted was the Tehran Convention of 2003, which entered into force in 2006 upon the ratification by all Caspian littoral states. Being a first, it was successful only in the areas of protection of the environment and guaranteeing of regional security. Another example of regional multilateral cooperation is the Caspian Security Agreement of 2010. The adoption of the above agreements indicated the possible direction of future actions by the coastal states aimed at seeking and reaching agreements on specific important aspects of their cooperation in the Caspian basin.

Aside from those two positive developments, there remained other aspects concerning the use of the Caspian Sea that required new developments—just to name the most urgent ones: navigation, fishing, resource extraction, and laying of Trans-Caspian pipelines. The best-case scenario would have been to include all these aspects as part of an overall convention on the status of the Caspian Sea. However, the separate regime-related agreements adopted before the conclusion of the final status agreement were not able to cover other contested fields in use of the Caspian Sea.

4.5.1 Protection of the Marine Environment of the Caspian Sea

Overexploitation, habitat destruction, and pollution threaten the natural resources of the Caspian Sea. There are also problems caused by water-level change, which greatly reduced fish stocks (especially sturgeon) in the Caspian. The introduction of alien fish species in the Volga–Don also poses a threat. In the face of a significant growth of concern regarding the poor environmental protection of the Caspian Sea, it has been necessary to take all appropriate measures to prevent the further deterioration of its ecosystem.

Since the breakup of the Soviet Union, there have been a lot of divergent views for solving the current legal challenges involving the Caspian Sea, including environmental protection. Until 2018, negotiations between the coastal states have proved to be successful only in regard to the issue of protection of the Caspian environment. The first attempt to establish a regional institutional framework among the Caspian Sea states to tackle environmental challenges was a proposal submitted by Iran in 1992 to set up a regional organization comprising the Caspian Sea states.[31] The limited success of this initiative can be attributed on the one side to the different political priorities of the littoral states, some of which were focused rather on the development of mineral resources, and on the other side to the lack of financial resources to address environmental issues. The first legal step toward the mutual protection of the Caspian environment was the adoption in 1994 of the Almaty Declaration on Cooperation of the Environmental Protection of the Caspian Sea Region. Respectively, regional cooperation on environmental methods grew, mainly thanks to the support of international partners like the United Nations Environment Programme (UNEP), United Nations Development Programme (UNDP), World Bank, Global Environment Facility (GEF) and EU's Technical Assistance for the Central Independent States (EU/TACIS). They supported the Caspian countries in the establishment of the so-called Caspian Environmental Programme (CEP)[32] in May 2018, with the goal of dealing with environmental issues and enhancing the framework for sustainable development in the Caspian region. The CEP helped states to establish the Convention, as well as regional and national action plans like the National Environmental Action Plan (NEAP), the Strategic Action Plan (SAP), as well as the Strategic Convention Action Program. These plans were implemented in the second phase of the CEP, which helped to develop a legally binding arrangement for the region, the Framework Convention for the Protection of the Marine Environment of the Caspian Sea (Tehran Convention). Overall, CEP activities facilitated the creation of favorable conditions for regional interstate cooperation on environmental issues.

---

[31] Blum (2002).
[32] UN Environment Programme—Caspian Sea <https://www.unenvironment.org/explore-topics/oceans-seas/what-we-do/working-regional-seas/regional-seas-programmes/caspian-sea> accessed July 5, 2020.

4.5 Step-by-Step Multilateral Regulations of the Legal Regimes in the Caspian Sea

At the end of the conference in Tehran in November 2003, the Caspian littoral states signed a Final Act, of which the Framework Convention for the Protection of the Marine Environment of the Caspian Sea (Tehran Convention) constitutes Annex 2. The Tehran Convention entered into force on August 12, 2006, after being accepted by all Caspian littoral states. At present, four additional protocols have been adopted—the Protocol Concerning Regional Preparedness, Response and Co-operation in Combating Oil Pollution Incidents (Aktau Protocol) (2011); the Caspian Countries Sign the Land-Based Sources and Activities Protocol (Moscow Protocol) (2012); the Biodiversity Protocol (2014); and the Environmental Impact Assessment (2018)—but only one of them has entered into force. The Aktau Protocol is the first and only ratified protocol to the Tehran Convention, which entered into force on July 25, 2016. The Moscow Protocol has been ratified by Azerbaijan, Iran, and Turkmenistan. The Biodiversity Protocol was ratified only by Turkmenistan, and the Environmental Impact Assessment was ratified only by Azerbaijan. Since 2014, a legal instrument (protocol) on monitoring assessment and information exchange has been under consideration, but it has not yet been adopted. Additionally, an Agreement on Hydrometeorology Cooperation in the Caspian Sea was adopted in 2014, with the goal of creating and developing an integrated regional system for receiving and exchanging information concerning the conditions in the Caspian Sea in the interest of ensuring the safety of life and the sustainable development of economic activity at the sea.

The Caspian Sea Convention, adopted in 2018, is scarce in provisions on ecology. It refers, however, to the provisions of the Tehran, which remains a main source of legal regulation for environmental protection in the Caspian Sea.

As the name suggests, the "Framework Convention for the Protection of the Marine Environment of the Caspian Sea" is aimed at the environmental protection of the Caspian Sea. The Tehran Convention (Article 4) includes states' general obligations to take individually or jointly all appropriate measures to prevent the pollution of the Caspian Sea and to protect the environment of the Caspian Sea. The Tehran Convention developed procedural regulations for the better implementation of the states' general commitments. These include environmental impact assessment, technological and scientific cooperation between the contracting parties, monitoring, and exchange of information and access thereto.[33] The Tehran Convention consists of internationally recognized principles necessary to achieve its objectives and implement its provisions. Also, regulations concerning the prevention, reduction, and control of pollution, as well as measures for the protection, preservation, and restoration of the marine environment, are part of the Convention. All

---

[33]Tehran Convention, Articles 17–21; Barcelona Convention 1986, Articles 10, 11, and 20; Kuwait Convention 1978, Articles X–XII and XXIII; Abidjan Convention 1981, Articles 13, 14 and 22; Lima Convention 1981, Articles 7–10 and 14; Jeddah Convention 1982, Articles X–XII and XXII; Cartagena Convention 1983, Articles 12, 13 and 22; Nairobi Convention 1985, Articles 13, 14 and 23; Noumea Convention 1986, Articles 16–19.

these provisions establish obligations aimed at the abatement of pollution from different sources: land-based sources, seabed activities, vessels, and dumping.[34]

The "framework" feature of the Tehran Convention is supposed to establish a template for the ongoing diplomatic process to reduce the pollution in the Caspian Sea caused by various sources. In comparison with similar international conventions and agreements, the provisions of the Tehran Convention are formulated in a rather vague way. The geographic boundaries of the Caspian are not clearly defined, and timelines are almost entirely absent from this Convention. The Tehran Convention does not name specific threats to the environment of the Sea, not even oil, which is the most dominant source of pollution. There is no direct reference to protected zones existing in the Caspian or to the threat of overfishing of sturgeon or other endemic species. There is no definition of the notion "rare and endangered species," nor are the "adequate emergency preparedness measures, adequate equipment, and qualified personnel" to respond to environmental emergencies defined. The goal of the coastal states for the Tehran Convention was to negotiate protocols on specific environmental issues,[35] which would define the environmental protection of the Caspian Sea in more detail. Up to present, significant work continues in the form of adopting additional protocols to the Tehran Convention, which shall operationalize its work. A serious weakness of the process of environmental law setting is that civil society organizations are not involved in it. Three of the five states parties, which are signatories to the Convention on Access to Information, Public Participation in Decision-Making and Access to Justice in Environmental Matters (further referred to as Aarhus Convention), are obliged to involve the public.

Taken as a whole, the Tehran Convention marks a step forward in the coastal states' effort to preserve the particularly fragile maritime environment of the Caspian Sea. The high complexity of the regulation in question resulted in a number of international partners becoming involved in the negotiation process preceding the adoption of the Tehran Convention, for instance the Caspian Environmental Programme, under the auspices of the United Nations Environment Programme, as well as the Global Environment Facility (GEF), a joint venture of the United Nations Development Programme (UNDP), the United Nations Environment Programme (UNEP), and the World Bank. The provisions of the Tehran Convention are linked to the standard rules and norms of current international law, referring to the agreements concluded both at international and regional levels. It has already been demonstrated by the UNEP Regional Sea Project and will be seen in the following analysis. The Tehran Convention can be classified as an example of regional regulations, which include treaties under the UNEP Regional Seas Programme and ad hoc regional and

---

[34] Tehran Convention, Articles 7–10; 1976 Barcelona Convention 1986, Articles 4–9; 1978 Kuwait Convention 1978, Articles II–IX; Abidjan Convention 1981, Articles 4–9 and 12; Lima Convention 1981, Articles 3–6; Jeddah Convention 1982, Articles III–IX; Cartagena Convention 1983, Articles 3–11; Nairobi Convention 1985, Articles 3–12; Noumea Convention 1986, Articles 4–9 and 15.

[35] Art. 6 (implementation); 7.2 (prevention, reduction and control of pollution); 8, 9, 10 (pollution); 14.2 (protection, preservation and restoration of marine biological resources); 18; 16 (sea level fluctuation); 17 (procedures of environmental impact assessment).

subregional arrangements for Europe and the Antarctic. As mentioned before, the preparations for the Framework Convention took place under the auspices of UNEP, which has clearly exerted influence on the approach between the parties. The UNEP Regional Seas Programme, launched after the 1972 Stockholm Conference and the creation of UNEP itself, was aimed at developing rules and norms at regional level[36] and now extends to 13 regional areas.[37] The eight regional sea framework conventions include substantive and procedural obligations, institutional arrangements, and provisions regarding the adoption of protocols and annexes. The same structure is featured in the Tehran Convention for the Caspian Sea.

The Tehran Convention reflects a worldwide larger trend toward greater international regulation of environmental protection. The recognition and protection of the environment lead, on the one hand, to a considerable restriction of state sovereignty and, on the other, to the recognition of the values by which all states are bounded, primarily the protection of the environment. However, the Tehran Convention explicitly states that none of its provisions "shall be interpreted as to prejudge the outcome of the negotiations on the final legal status of the Caspian Sea" (Article 37). Many references to the global and regional agreements that have built the legal basis for the Tehran Convention, including provisions typical for seas as well as for international watercourses, neither refer to the status of the Caspian Sea nor disclose states' official position on its status, which was finally defined in 2018 in the Caspian Status Convention.

A detailed examination of the significant lawmaking role of the Tehran Convention in the protection of the marine environment of the Caspian Sea has been presented in a separate chapter of this book. It presents a rather practical approach to the examining the Tehran Convention, based mostly on an analysis and a comparison of related international treaties and agreements. The adequacy of the Tehran Convention was judged by its ability to protect the marine environment of the Caspian Sea.

4.5.2 Aquatic Biological Resources of the Caspian Sea

During the Fourth Summit of the Caspian Sea heads of states on September 29, 2014, one more legal document related to the environmental conditions of the Caspian Sea was concluded. The Agreement on Conservation and Rational Use of the Aquatic Biological Resources of the Caspian Sea applies to aquatic biological resources, which cover fish, shellfish, crustaceans, mammals, and other species of aquatic animals and plants (Article 2). This Agreement aims at the conservation and rational use of the aquatic biological resources of the Caspian Sea, including the management shared aquatic biological resources (Article 3). This Agreement

---

[36]Mediterranean Action Plan (1975), p. 481.
[37]See: (Sands 1995), pp. 296–302.

foresees a set of principles, as well as mechanisms of an able state to implement its requirements. The states agreed to develop a research study, collect data, and exchange scientific-technical documents, experience, and information. The Agreement requires the states to take prevention measures for illegal, unreported, and unregulated fishing and to develop programs for aquatic production and for their reintroduction into natural habitats. The parties shall establish and join the committee. The Agreement was approved merrily by Azerbaijan and entered into force on May 24, 2016. A more specific overview of this Convention will be provided in the Chap. 7.

4.5.3 Cooperation on the Prevention and Elimination of Emergency Situations in the Caspian Sea

In 2014, all Caspian coastal states adopted the Agreement on Cooperation in the field of prevention and elimination of emergency situations in the Caspian Sea, which entered into force on September 19, 2017. This Agreement regulates the interaction between the Caspian states in case of natural and man-made disasters in the Caspian Sea (Article 2). The goal of this Agreement is the prevention and/or elimination of emergency situations in the Caspian Sea, and in case of such, a party shall be entitled to apply for assistance from other states. An emergency situation, in accordance with the Agreement, means a situation resulting from *an accident, disaster or disaster of a natural or man-made nature, which has resulted in, or may result in, loss of human life, damage to human health, the environment and facilities of the production and social infrastructure, significant material losses and violations of the living conditions of the population* (Article 1). Parties of this Agreement shall create and improve the system of interaction on prevention and liquidation of emergency situations, forecasting and monitoring of emergency situations, provide information about the risks of such situation, offering emergency relive, sharing experience, exchanging information and holding common conferences and other activities. The Agreement regulates the procedures to request assistance (Article 6), coordination and cooperation in emergency situations (Article 7), as well as offers provisions for transit, entry, and exit of the state in the territory of the Party requesting aid, conditions for using of air and water transport, and regulating costs of assistance. More details on this Agreement will be provided in Chap. 10.

4.5.4 Security Cooperation in the Caspian Sea

Poaching, illegal immigration, arm trafficking, drug trafficking, and organized crime are classified as illegal acts under international law. At present state, these problems are intensified in the Caspian basin, representing a significant challenge for the

4.5 Step-by-Step Multilateral Regulations of the Legal Regimes in the Caspian Sea

coastal states.[38] Not only the safety of maritime transport but also the environment of the Caspian Sea can be threatened by maritime terrorism and piracy. Therefore, the states strengthen their cooperation to fight against these threats and also in working out and implementing relevant legal measures. The duty to cooperate in the repression of illegal acts at sea remains primarily important for coastal states.

The legal framework for combating security risks in the Caspian Sea was defined in the Agreement on Security Cooperation in the Caspian Sea (further referred to as Caspian Security Agreement). It was concluded in Baku on November 18, 2010, which entered into force on September 27, 2014. It reaffirms the commitment of the parties to contribute to regional security and stability, development, and the strengthening of cooperation in the use of the Caspian Sea exclusively for peaceful purposes. It states that security in the Caspian Sea is the prerogative of the littoral states. At the same time, nothing in this Agreement is intended to prejudice the future shape of the legal status of the Caspian Sea, in respect of which international negotiations are still underway.

The Caspian Security Agreement contains forms of cooperation of the competent authorities of the parties (Article 2)—procedures for the exchange of information and experience, meetings, consultation, and concerted action to address the fight against terrorism, organized crime, illegal arms dealing, trafficking of drugs, money laundering, smuggling, piracy, human trafficking, illegal migration, and illegal exploitation of biological resources (poaching) and to ensure the safety of navigation. To solve any security-related problems, countries hold meetings and consultations as necessary, but not less frequently than once a year.

According to UNCLOS, states are entitled—on the high seas or in any other place outside the jurisdiction of any state—to repress piracy (Article 100), suppress illegal dealings in drugs or psychotropic substances (Article 108), cooperate in the suppression of unauthorized broadcasting from the high seas (Article 109), etc. However, effective marine policing is missing in the Caspian Sea. The Caspian Security Agreement does not provide for its establishment either.

During the Fifth Summit of the Caspian states in 2018, there were three additional protocols that were adopted into the Caspian Security Agreement. First, the Protocol about Cooperation in the area of the fight against organized crime on the Caspian promotes cooperation among the states for the prevention, detection, suppression, and disclosure of crimes committed in organized forms (Article 1). The Protocol lists the relevant state agencies that should cooperate in the fight against crimes against persons like human trafficking, economical crimes like money laundering, illegal migration, etc. For these purposes, the relevant agencies shall exchange information, conduct common activities, and exchange experiences (Article 6). The cooperation shall be upon the request of one of the parties (Article 4).

Second, the Protocol About Cooperation in Fight Against Terrorism provides that countries should cooperate in the fight against terrorism according to international and national laws through the exchange of information, meetings and consultations,

---

[38] See: (Musing 2019).

and coordinated actions for the prevention, detection, and suppression of activities of a terroristic character (Article 3).

Third, the Protocol about Cooperation and Interaction of Border Agencies shall be applied in the areas defined in Article 2 of the Caspian Security Agreement: the fight against terrorism; the fight against organized crime; the fight against illicit trafficking in weapons of any kind, ammunition, and explosive and toxic substances of military equipment; the fight against illicit trafficking in narcotic drugs, psychotropic substances, and their precursors; the fight against money laundering, including of funds obtained by criminal means; the fight against smuggling; ensuring the safety of maritime navigation and the fight against piracy; combating trafficking in persons and illegal migration; the fight against the illegal extraction of biological resources (poaching); and ensuring the safety of navigation. The Protocol lists relevant agencies on the central and regional levels that are supposed to exchange information and hold seminars for the exchange of experiences, as well as conduct activities for the prevention, detection, and suppression of illegal activities in the Caspian Sea (Article 3).

4.6 Current Legal Framework for the Status of the Caspian Sea as Defined in the Caspian Sea of 2018

The adoption of the Caspian Sea Convention on August 12, 2018, is to be seen as a milestone in the regulation of the Caspian Sea, which could strengthen the interstate use of the Caspian Sea. For almost 30 years after the Soviet Union's collapse and the emergence of five independent states (Azerbaijan, Kazakhstan, Russia, Turkmenistan, and Iran), the legal status of the Caspian Sea had not been well defined.

The Caspian Sea Convention defines the Caspian Sea as a "*'body of water'* (emphasis added) surrounded by the land territories of the Parties" (Article 1). The acceptance of such a legal definition has allowed the negotiating states to develop an independent set of rules regulating the extent of state sovereignty over the Caspian Sea, including the development of natural resources, maritime transport, laying pipelines, and any other activities on the Caspian Sea water column, as well as on its seabed and subsoil. Although, as the reading of its text shows, the Caspian Sea Convention follows the legal framework prescribed in UNCLOS,[39] still one may see deviations from UNCLOS in the regulation of the Caspian Sea's status and related forms of usage. The difference between the UNCLOS regulations and the Caspian Sea Convention can be traced from the fact that Caspian riparian states wanted to develop a specially tailored legal status for the maritime zones of the Caspian Sea. Since there has never been a mutually recognized position that the Caspian Sea is a sea in legal terms and, respectively, that it is obligatory for the riparian states to follow the Law of the Sea provisions when defining status of the Caspian Sea, the

---

[39]UN Convention on the Law of the Sea (UNCLOS), Dec. 10, 1982, 1833 U.N.T.S. 397.

parties to the Caspian Sea Convention have developed a unique legal setting that best reflects their current political positions.

The scope of the Caspian Sea Convention is defined in its Article 2, which states:

> In accordance with this Convention, the Parties shall exercise their sovereignty, sovereign and exclusive rights, as well as jurisdiction in the Caspian Sea.

The goal of the Convention is to define *the rights and obligations of the Parties in respect of the use of the Caspian Sea, including its waters, seabed, subsoil, natural resources and the airspace over the Sea* (Article 2(2)).

According to Article 4, *the Parties shall conduct their activities in the Caspian Sea for the purposes of navigation, harvesting, use and protection of aquatic biological resources, exploration and exploitation of the seabed and subsoil resources, as well as other activities in accordance with this Convention, other agreements between the Parties consistent with this Convention, and their national legislation.*

In addition, the Preamble also reflects the overall goals of the Caspian Sea Convention. In the Preamble, the states parties express their desire to be guided toward deep and good neighborly relations. They want to take account of the atmosphere of cooperation, good neighborliness, and mutual understanding among them. States are ready to recognize the political, economic, social, and cultural importance of the Caspian Sea and their responsibility to ensure its sustainable development for the present and future generations. Parties are aimed at a peaceful and rational management of Caspian natural resources, as well as the exploration, protection, and conservation of its environment since, as they reconfirm merely, they possess sovereign rights over the Caspian Sea and its resources. Any arising issues will be solved within the exclusive competence of the parties to create favorable conditions for the development of a mutually beneficial economic cooperation in the Caspian Sea. The Preamble reconfirms that the Caspian Sea Convention is based on the principles and norms of the Charter of the United Nations and the International Law, as well as takes into account existing arrangements between the parties.

4.6.1 General Principles for Interstate Cooperation

The activities of the coastal states shall be conducted based on the principles defined in Article 3 of the Caspian Sea Convention. The overall principles are

> the respect for the sovereignty, territorial integrity, independence and sovereign equality of States, non-use of force or the threat of force, mutual respect, cooperation and non-interference into the internal affairs of each other (Article 3, Para. 1).

The Caspian Sea shall be used

> for peaceful purposes, making it a zone of peace, good-neighborliness, friendship and cooperation, and solving all issues related to the Caspian Sea through peaceful means (Article 3, Para. 2).

In terms of regional security and stability in the region, states shall ensure

a stable balance of armaments of the Parties in the Caspian Sea, developing military capabilities within the limits of reasonable sufficiency with due regard to the interests of all the Parties and without prejudice to the security of each other (Article 3, Para. 4)

Compliance with the agreed confidence-building measures in the military field in the spirit of predictability and transparency in line with general efforts to strengthen regional security and stability, including in accordance with international treaties concluded among all the Parties (Article 3, Para. 5)

Non-presence in the Caspian Sea of armed forces not belonging to the Parties; (Article 3, Para. 6)

Non-provision by a Party of its territory to other States to commit aggression and undertake other military actions against any Part (Article, 3 Para. 7)

For guaranteeing the freedom of navigation in the Caspian Sea outside of the territorial waters of the states, they have agreed on the following:

Freedom of navigation outside the Territorial Waters of each Party subject to the respect for sovereign and exclusive rights of the coastal States and to the compliance with relevant rules established by them with regard to the activities specified by the Parties; (Article 3, Para. 8)

Ensuring safety of navigation; (Article 3, Para. 9)

The right to free access from the Caspian Sea to other seas and the Ocean, and back in accordance with the generally recognized principles and norms of international law and agreements between the relevant Parties, with due regard to legitimate interests of the transit Party, with a view to promoting international trade and economic development; (Article 3, Para. 10)

Navigation in, entry to and exit from the Caspian Sea exclusively by ships flying the flag of one of the Parties; (Article 3, Para. 11)

The Caspian Sea Convention guarantees the freedom of aviation, saying that

freedom of overflight by civil aircraft in accordance with the rules of the International Civil Aviation Organization (Article 3, Para. 16).

The protection of the environment of the Caspian Sea foreseen in the Caspian Sea Convention includes its

conservation, restoration and rational use of its biological resources has been confirmed (Article 3, Para. 14).

The shared aquatic biological resources shall be reproduced and regulated according to agreed norms and rules. The Caspian Sea Convention states the

liability of the polluting Party for damage caused to the ecological system of the Caspian Sea (Article 3, Para. 13).

Ecology and conservation of biological resources shall be subject to scientific research. The overall conduct of maritime scientific research outside of the territorial waters of the parties will be guaranteed.

There is no reference in Article 3 to the general principles governing the use of the nonliving resources on the Caspian Sea, apart from the general principle for the respect of state sovereignty, which in the Caspian Sea sectors will govern the use of nonliving resources.

4.6.2 Security Cooperation According to the Caspian Sea Convention

The Caspian Sea Convention reconfirms the provisions included in the Caspian Security Agreement and states:

> The Parties shall cooperate in combating international terrorism and financing thereof, trafficking in arms, drugs, psychotropic substances and their precursors, as well as poaching, and in preventing and suppressing smuggling of migrants by sea and other crimes in the Caspian Sea (Article 17).

The overall duty of the Caspian Sea states is to carry out their activities in the Caspian Sea in accordance with the principles of ensuring the following:

- Ensuring safety of navigation (Article 3 Para. 9)
- Ensuring security and stability in the Caspian Sea region (Art. 3 Para. 3)
- Ensuring a stable balance of armaments of the Parties in the Caspian Sea, developing military capabilities within the limits of reasonable sufficiency with due regard to the interests of all the Parties and without prejudice to the security of each other; (Art. 3 Para. 4)
- Compliance with the agreed confidence-building measures in the military field in the spirit of predictability and transparency in line with general efforts to strengthen regional security and stability, including in accordance with international treaties concluded among all the Parties; (Art. 3 Para. 5)
- Non-presence in the Caspian Sea of armed forces not belonging to the Parties; (Art. 3 Para. 6)
- Non-provision by a Party of its territory to other States to commit aggression and undertake other military actions against any Party; (Art. 3 Para. 7)

The Caspian Sea Convention guarantees the legal regime of warships in the Caspian Sea, which will be presented in detail in Chap. 9.

4.7 Conclusion

The dissolution of the Soviet Union sharpened the conflicting interests in the use of the Caspian Sea, especially of its living (sturgeon, etc.) and nonliving (oil and gas) natural resources. The unclear legal status of the Caspian Sea, resulting from the Soviet–Iranian treaties as well as regional state practice, has pushed states to an active search for legal solutions. Approaches adopted by the states were multidimensional. First, there was the multilateral approach that involved all coastal states. It was the most sustainable way of resolving the existing conflicts around the Caspian Sea, which ended successfully with the conclusion of the Caspian Sea

Convention in 2018. The initial form of multilateral cooperation between the coastal states was entering multilateral negotiations for a comprehensive solution to the question of legal status, an output of which was the preparation of a Draft Convention on the Legal Status of the Caspian Sea. This document regulates all aspects of the legal status and regime of the Caspian Sea. Its complexity prevented states from successfully finalizing the negotiations until 2018, which started in the mid-1990s. Its added value is that this document, even though not in force, offered quite an exact picture of the legal positions of the negotiating states and therefore allowed to draw some conclusions regarding the future legal developments to be expected in the region. However, its lacking of any legal force pushed some coastal states toward other parallel solutions. Chronologically, the first bilateral legal actions were taken by Russia, Kazakhstan, and Azerbaijan, aiming at the division of the North Caspian seabed and subsoil for the use of its natural resources. These actions brought about great disagreement among the remaining states. Acts of a multilateral character concerning the Caspian Sea were as follows: first, the adoption of the Tehran Convention, where coastal states agreed upon the mutual protection of the Caspian environment in 2013, and, second, states' concluding of an agreement guaranteeing regional security (the Caspian Security Agreement in 2010 and the Agreement on the Conservation and Rational Use of Aquatic Biological Resources of the Caspian Sea of 2014).

Even though all states parties reassured that neither bilateral nor sectoral multilateral approaches hamper prospects for the future resolution of the dispute around the status of the Caspian Sea, the over 20 years of multilateral negotiations proved to be very complicated. For some time, it seemed rather realistic that multilateral sectoral regulations will be an effective mechanism capable of resolving the existing conflict around the Caspian Sea. Obviously, sectoral multilateral solutions were less contentious than bilateral agreements, which excluded other coastal states. There were not many sectoral issues, apart of security and bio resources, which were mutually supported by all Caspian Sea states. Still, the case of sectoral agreements did not offer a long-term solution to the question of legal delimitation of the Caspian Sea or a clarification of the scope of the states' sovereignty over the Caspian Sea. It was resolved only through the adoption of the Convention on the Caspian Sea Legal Status in 2018, which has opened a new area of interstate cooperation, building up a firm legal framework for the development of the region.

References

Bernhardt R (1984) Einfluß der UN-Seerechtskonvention auf das geltende und zukünftige internationale Seerecht. In: Delbrück J (ed) Das neue Seerecht. Duncker & Humblot, Berlin

Blum D (2002) Beyond Reciprocity: Gouvernance and Cooperation around the Caspian Sea. In: Conca K, Dabelko G (eds) Environmental peacemaking. Woodrow Wilson Center Press, p 161 (б.м.)

References

Caron D, Shinkaretskaya G (1995) Peaceful settlement of disputes through the rule of law. In: Damrosch LF, Danilenko GM, Mullerson R (eds) Beyond confrontation. International law for the post-cold war era. Westview, San Francisco

Gornig G, Despeux G (2002) Seeabgrenzungsrecht in Der Ostsee: Eine Darstellung Des Voelkerrechtlichen Seeabgrenzungsrechts Unter Besonderer Berucksichtigung Der Praxis Der Ostseestaaten. Peter Lang AG, Bern, Switzerland

Ibp I (2018) Kazakhstan Diplomatic Handbook – Strategic Information and Developments. IBP USA (б.м.)

Kimminich O (1997) Einführung in das Völkerrecht. 6th ред. Tübingen. Francke, Basel

Kunz J (1968) The changing law of nations: essays on international law. Ohio State University Press, Columbus

Land K (2000) Souveränität und friedliche Streitbeilegung. Europäische Hochschulschriften. 2950s ред. Peter Lang (б.м.)

Mamedov R (2001) International legal satus of the Caspian Sea; issues of theory and practice. Turkish Yearb Int Relat 32:217–259

Musing L (2019) Corruption and wildlife crime: a focus on caviar trade. TRAFFIC, Cambridge

Ndiaye T, Wolfrum R (2007) Law of the sea, environmental law and settlement of disputes. Koninklijke Brill NV, Leiden/Boston

North Sea Continental Shelf Judgement (1969) Federal Republic of Germany/Denmark

RadioFreeEurope (2002) Caspian: Ashgabat Summit Ends Without Agreement. [В Интернете] Available at: https://www.rferl.org/a/1099503.html [Дата обращения: 2 July 2020]

RadioFreeEurope (2007) Caspian: Summit Fails to Resolve Key Question. [Online] Available at: https://www.rferl.org/a/1078963.html. Accessed 2 July 2020

RadioFreeEurope (2010) Caspian Sea States Gather In Baku For Summit. [Online] Available at: https://www.rferl.org/a/Caspian_Sea_States_Gather_In_Baku_For_Summit/2223479.html. Accessed 2 July 2020

RadioFreeEurope (2014) Putin Optimistic Of Eventual Accord After Fourth Caspian Summit. [В Интернете] Available at: https://www.rferl.org/a/caspian-sea-summit/26610935.html [Дата обращения: 2 July 2020]

RadioFreeEurope (2018) Five States Sign Convention On Caspian Legal Status. [В Интернете] Available at: https://www.rferl.org/a/russia-iran-azerbaijan-kazakhstan-turkmenistan-caspian-sea-summit/29428300.html [Дата обращения: 2 July 2020]

Sands P (1995) Principles of international environmental law. Manchester University Press, Manchester

Sohn L (1950) Cases and other materials on world law. Foundation Press, Brooklyn (ред. б.м)

Chapter 5
Interrelations Between Territorial Delimitation and the Regime of the Use of the Caspian Sea

5.1 Nonlegal Aspects of Settlement of the Seaward Boundaries in the Caspian Sea

Historically, states used to be more interested in the questions of usage regime than setting maritime borders. The newly observed praxis is states' attempt to redefine the seaward extension of their sovereignty due to availability of modern technologies allowing for bigger than traditional exploitation of natural resources. The scarcity of world resources has resulted in states competing to expand their zones of influence with no limits and put forward claims concerning the delimitation of seaward boundaries.[1] Unresolved border disputes often lead to poor management of resources and of ecology and to disagreement and conflicts in interstate relations. This has been the situation in the Caspian Sea region since the collapse of the Soviet Union until recently.

Maritime delimitation requires regulation of the law of the sea that defines the legality of unilateral acts of states.[2] Mutually agreed intergovernmental boundary treaties are of great importance for the successful settlement of maritime disputes. Only the absence of relevant treaties authorizes direct reference to the norms of the law of the sea. However, it cannot be denied that the final lines of state maritime borders are determined not alone by law but to a great extent by nonlegal factors like political, historical, security, economic, environmental, and geographical circumstances as well. States are often not willing to admit the influence of these factors on their border setting. Nevertheless, to successfully deal with the issues of border setting and, relevantly, the regime of resource usage—including in the case of the Caspian Sea—one needs to also refer to the nonlegal conditions of delimitation.

International law does not determine the actual boundary lines. These are left to the political will of the neighboring states and their nonnormative interactions, such as security, foreign policy objectives, etc. Decision-making processes in the

---

[1] Boggs (1951).
[2] ICJ Decision in case Gulf of Maine, In: ICJ Rep. 1984, § 112, p. 299.

delimitation cases show similar policy-, economy-, ecology-, etc. related tendencies[3] worldwide. The political aspects of defining boundaries are regulated in intergovernmental negotiations, which begin from undertaking negotiations, defining negotiating positions, joint acceptance of proposed demarcation lines, etc. This experience has also been repeated in boundary setting negotiations in the Caspian Sea, which were carried on after the collapse of the Soviet Union until 2018, when the Caspian Sea Convention was signed finally.

Sometimes to retain flexibility in their relations, states avoid a clear formulation of provisions of boundary agreements. This approach seemed—for nearly 20 years of intergovernmental negotiations—to characterize the legal debate on the status of the Caspian Sea. How else could the slow progress in multilateral interstate negotiations on the agreement on the future status of the Caspian Sea until 2018 be explained? The unclear legal heritage of the Soviet Union in relation to Caspian borders and the use of its resources lasted despite the ongoing negotiations conducted since the early 1990s. The limited progress was to be attributed to the deep differences in geopolitical and economic interests of all the five littoral states that they were not ready to give up. The application of the middle-line principle for the eventual division of the Caspian Sea, which was demanded by the newly independent Caspian states, would award the traditional regional powers—Russia and Iran—merely a small part of the Caspian's surface and its resources, which in turn brought about a politically and economically motivated conflict between groups of states. The majority of interstate delimitation treaties worldwide have unilateral rather than multilateral character. Agreements regulating transboundary areas of cooperation, such as the conservation of highly migratory species, are an exception. The tendency for unilateral actions is present also in the Caspian Sea. Difficulties in the conduct of multilateral negotiations on the delimitation of the seabed of the entire Caspian basin cause coastal states to carry out unilateral actions. The most prominent case was the division of the North Caspian Sea between Azerbaijan, Kazakhstan, and Russia.

It can be observed in border negotiations worldwide that states carry out delimitation merely up to the starting point of a disputed territory, without hampering future delimitation of the remaining area.[4] Similar approach was taken for the delimitation of the Northern Caspian sectors between Azerbaijan, Kazakhstan, and Russia, and later Turkmenistan (North Caspian Agreements). They covered merely these sectors, which lay between the coasts of the states, and did not touch upon areas claimed by the remaining coastal states. It remained, however, controversial whether these treaties do not affect the interests of Turkmenistan and Iran regarding the common use of the Caspian Sea. However, except protests originally raised by both countries through diplomatic channels, they did not undertake any legal actions to secure their claims regarding the northern part of the Caspian Sea, as discussed earlier (Sect. 4.4).

---

[3]Oxman (1993), pp. 3 et seq.
[4]Charney and Alexander (2003), pp. 1057 et seq. No. 5–12.

The coastal states' claims regarding fishing stocks, exploration and exploitation of energy resources, or navigation reflect economic and security-related aspects of a marine delimitation. Such boundaries described as "historical borders" in international practice play a major political and legal role.[5] It happens that such informal or de facto borderlines turn into an official interstate border.[6] The natural resources already known or readily ascertainable in the areas under delimitation might well constitute relevant circumstances that would be reasonably considered during the delimitation process.[7] Some solutions to disputes regarding the delimitation of marine boundaries were largely affected by traditional claims of coastal states regarding the use of fishing resources. A close relationship between the boundary line and the traditional fishing rights was pointed out by the International Court of Justice in the case of *Anglo-Norwegian Fisheries*.[8]

In the unclear framework regarding the sharing of the Caspian Sea, the coastal states' policy regarding its delimitation was influenced by similar factors until 2018 and the conclusion of the Caspian Sea Convention. Fish stocks were traditionally the subject of joint use of the coastal states in the Caspian region. The seabed oil and gas resources existing in the Caspian Sea, as they form an indispensable basis for the independent economic and political existence of the coastal states, were claimed to be divided among the coastal states. Also, navigation is a globally recognized impact factor upon the delimitation of boundaries.[9] It was, however, directly applied only in a few cases, where navigation was recognized as "special circumstances" and it was determined how a border-line dispute was concluded.[10] The control or even exclusion of foreign vessels from the immediate vicinity of national coasts is regarded as a guarantee to strengthen national trade. This approach was present in all previously adopted regulations of the Caspian Sea, starting with the Soviet–Iranian treaties of the 1920s up to the present negotiations over the status of the Caspian Sea. Shipping in the Caspian is traditionally exercised on an equal footing by all coastal states, except free navigation by states outside of the region.

Intergovernmental practice shows that environmental factors, despite their relevance for the sustainable development of maritime areas, play a relatively minor role in the delimitation of borders. This was expressed in the case of the *Gulf of Main* decision of the International Court of Justice, which stated that boundaries cannot be determined by environmental factors because they cannot be regarded as

---

[5]Convention on the Territorial Sea and the Contiguous Zone 1958, Article 12; Convention on the Continental Shelf 1958, Article 6; UNCLOS, Article 15.
[6]Charney and Alexander (2003), pp. 1475 et seq. No. 7–1.
[7]Libya/Malta 1985, ICJ 4, Para. 50. Exception: Jan Mayen case, ibid., pp. 1755 et seq. No. 9–4.
[8]ICJ 1951, 133 and 142.
[9]Anglo-French Continental shelf case, in: ibid, pp. 1735 et seq. No. 9–3.
[10]Argentina–Chile, ibid., pp. 719 et seq. No. 3–1.

circumstances with catastrophic repercussions.[11] Awareness among the Caspian littoral states of the lack of clear regulations for the protection of the environment and the resulting extensive damage to the fragile Caspian environment persuaded the littoral states to adopt in 2003 the Tehran Convention on the environmental protection of the Caspian Sea but did not solve the overall negotiation issue regarding Caspian Sea legal delimitation.

Also, the pressure from external political powers, which have their own economic and security interests in resource-rich regions worldwide, impacted the conduct of maritime delimitation there. This phenomenon can also be observed in the politics in the Caspian region. The global powers are actively involved in Caspian affairs, indirectly influencing regional delimitation. The first one to name is the United States, active by signing the "Contract of the Century" and through the construction of the Baku–Ceyhan pipeline and more.[12] Also, the recently announced EU Commission's 2019 Central Asia strategy states that the EU will continue to focus on enhancing the role of Central Asia in contributing to energy supply security and the diversification of suppliers, sources, and routes of the EU, including assessing the possibility to build the Trans-Caspian pipeline.[13] A recent initiative launched by China, "One Belt, One Road," highlights the significant role of the Caspian Sea in international trade.[14] The Trans-Caspian International Transport Route (TITR, also known as the "middle corridor"), which includes 4256 km of railways and 508 km of maritime transit, stretches from the Chinese–Kazakh border to Europe and crosses the Caspian Sea through Kazakhstan via Azerbaijan and Georgia and the possibility of becoming a major multimodal artery within the developing commercial transport network of Eurasia.[15]

The conclusion of bilateral and trilateral delimitation agreements in the northern Caspian Sea brought some legal stability to the region but also defused the urgency for an overall solution, hampering the process of multilateral negotiations over the demarcation of the entire Caspian Sea. The Caspian Sea Convention concluded in 2018 has finally settled the lines of the state maritime borders. The new legal standards reflect and balance the dependence on nonlegal factors like political interests of the coastal states and external powers, economic development in the area of living and nonliving resources, as well as environmental conditions.

---

[11] 1984 ICJ 341, Para. 233.

[12] Rahim Rahimov, 'Prospects for the Trans-Caspian Gas Pipeline Under the Trump Administration' (July 18, 2019) Wilson Center https://www.wilsoncenter.org/blog-post/prospects-for-the-trans-caspian-gas-pipeline-under-the-trump-administration accessed July 15, 2020.

[13] European Commission and High Representative of the Union for Foreign Affairs and Security Policy, "The EU and Central Asia: New Opportunities for a Stronger Partnership," May 15, 2019, https://eeas.europa.eu/sites/eeas/files/joint_communication_-_the_eu_and_central_asia_-_new_opportunities_for_a_stronger_partnership.pdf accessed July 2, 2020.

[14] Tristan Kenderdine, 'Caucasus Trans-Caspian Trade Route to Open China Import Markets' (East Asia Forum, Feb. 23, 2018) https://www.eastasiaforum.org/2018/02/23/caucasus-trans-caspian-trade-route-to-open-china-import-markets/ accessed July 2, 2020.

[15] John Calabrese "Setting the Middle Corridor on track" (Middle East Institute, Nov. 18, 2019) https://www.mei.edu/publications/setting-middle-corridor-track accessed July 2, 2020.

5.2 Territorial Delimitation and State Sovereignty

With the collapse of the Soviet Union and the emergence of five independent states in the region of the Caspian Sea, which claimed full sovereignty or limited sovereignty rights (the so-called functional rights) in relevant sectors of the Caspian Sea, the question of delimitation of the Caspian Sea has become urgent. The lack of well-established national maritime zones often leads to poor resource and environmental management,[16] as well as to discord, conflicts, or even border disputes in interstate relations.[17]

International practice in boundary setting provides a reference to assess the course of states negotiations in the Caspian Sea. First, the geographical area to be delimited should have been identified, and only then could the delimitation process be launched. The definition of the territorial scope contributes to the protection of rights of third countries. However, the geographical definition of the so-called Caspian region poses difficulties.[18] Experts are not in agreement as to whether the Caspian Sea reflects the geographical characteristics of a "sea." For legal purposes, the subject of examination should be limited to a geographically, clearly fixed water area defined by the land coasts of five states, namely Azerbaijan, Iran, Kazakhstan, Russia, and Turkmenistan.

Second, the spatial delimitation of maritime zones remains closely linked to the amount of state sovereignty in such zones, which can take the form of full sovereignty of the littoral states or states' sovereign rights (the so-called functional rights). According to the law of the sea, every coastal state may exercise rights with regard to the use of living or nonliving resources located in its national maritime zone. The scope of coastal states' rights and the extension of the zones should be clearly defined between coastal states. According to the law of the sea, interstate agreements on the delimitation of borders are based on global[19] or regional law of treaties and customs as well as courts' and arbitration decisions. As the delimitation of maritime borders follows the rules of the law of the sea, it should require that the agreement between the Caspian Sea coastal states matches with these international standards. These shall be initially investigated on to determine what conditions must be fulfilled for the Caspian Sea's delimitation.

Delimitation is a legal act that aims at separating two sovereign or functional maritime areas. It should not be confused with demarcation, which is an act of technical character and intends to mark a line. A claim to conduct maritime

---

[16]Caspian Environmental Information Center. Caspian Sea 2018 State of Environment Report. https://ceic-portal.net/fa/node/3173 accessed July 2, 2020.

[17]Orazgaliyev and Araral (2019).

[18]Zonn and Zhiltsov (2004), pp. 7 et seq.

[19]Convention on the Territorial Sea and the Contiguous Zone of 1958; Convention on the Continental Shelf of 1958; Convention on the High Seas 1958; Convention on Fishing and Conservation of Living Resources of the High Seas; UNCLOS.

delimitation is only justified if states meet certain conditions. Only a state having a coast[20] and exercising sovereignty over an adjacent territory may lawfully claim its rights over maritime zones and enjoy a title to establish such zones.[21] State sovereignty thus provides a competence title regarding areas to be delimited.[22] It is part of the concept of the so-called natural prolongation, which explains that a state's maritime boundaries should reflect the "natural prolongation" of where its land territory meets the coast.[23] When claiming relevant maritime zones, states must be able to demonstrate the necessary legal title with regard to the boundary. National maritime zones seaward from land boundaries, which do not exceed international maritime borders, can be claimed by states by unilateral acts. Other conditions of maritime delimitation are that it has to be conducted in a zone where at least two legitimate titles overlap and with consideration of the existing and future rights of third states. The existence of these conditions in the case of the Caspian Sea will be examined further.

For the case of delimitation of the Caspian Sea, there were three legal methods available. The first method of the so-called tripoints was applied in regional legal practice for the peaceful settlement of the delimitation dispute in the North Caspian area, where the maritime areas of the three coastal countries converge and overlap.

The second one, so-called principle of "res inter alios acta,"[24] did not lead to an effective solution since some coastal states (in case of the Caspian Sea—three of them) have already concluded bilateral agreements without paying attention to the opposition of third countries.[25] In such case, no state can claim rights over the areas received, thanks to such a delimitation treaty, unless this deal does not harm the rights of third states. The North Caspian Agreements were concluded without taking

---

[20]Judgement Gulf of Maine, In: ICJ Rep. 1984, p. 296 § 102; Judgement Tunisia vs. Libya, In: ICJ Rep. 1982, p. 61 § 73.

[21]Gornig and Despeux (2002), pp. 28 et seq.

[22]On the departure from the principle of absolute sovereignty in international law, the judgment of the ICJ in "Trail Smelter case," in which the Court clarified the example damaged U.S. American Agriculture and Forestry caused by Canadian emissions, stating that in the event of a significant harm impairment of the principle of absolute sovereignty of a state could be abandoned and its sovereignty restricted. In: 162 LNTS 73, 3 RIAA 1907, 1938.

[23]ICJ judgment in case North Sea Continental Shelf, ICJ Reports 1969, pp. 3–32.

[24]The principle "*res inter alios acta*" means that a boundary between two States has no binding effect on a third country, and a modification of this principle is the principle "*res inter alios judicata aliis necnocere potest*" (see: Article 59 of the ICJ Statute: The decision of the Court has no binding force except between the parties and in respect to that particular case). This principle has the consequence that in the delimitation of an overlapping zones neither future nor existing claims of third countries are to be considered. This principle was applied by Courts of Arbitration in the cases against France, the United Kingdom (RSA, Vol XVIII, § 25, p. 154) and the US-American cases (Report on the New Jersey Delaware Maryland CEIP Delimitation Lines, pp. 23 and 27–28: Mississippi against Louisiana, Maryland against Delaware and Delaware against New Jersey).

[25]Such a situation was dealt with in the judgment of the ICJ regarding the delimitation of the North Sea continental shelf, where the States have concluded each new bilateral delimitation agreements without regard to previously existing contracts, See: ICJ Rep. 1969, § 4, pp. 13 et seq.

into consideration the territorial claims of Iran and Turkmenistan, which asserted in official statements that the Caspian Sea may be divided merely upon the consent of all the coastal states.

The application of the third available method of delimitation, which provides for the recognition of third countries' rights,[26] would result in disproportionate claims of some countries in the Caspian Sea. Iran claims a need for the equal delimitation of the Caspian Sea, where each coastal state receives a share of 20%. As the Iranian coast is the shortest one, it does not seem possible that other states would recognize its claims to a share of 20%. The application of the principle of recognizing third states' rights as a method of delimitation raises doubts as to whether it is right to expand the influence of third countries to such an extent that the assertion of claims to certain zones by coastal states depends entirely on the demands of third countries.[27]

The emergence of independent and sovereign states bordering the Caspian Sea has caused the need for the delimitation of maritime zones. Both the remains of the deadlocked regional Soviet practice and the intergovernmental cooperation ongoing for the last 20 years indicated the possible application of different boundary concepts in the Caspian Sea. Their brief assessment appears necessary to better portray the legal consequences of the application of any of them for the use of the Caspian resources. The final solution has been reached in 2018, when the Caspian Sea Convention was signed, which defined the delimitation course in the Caspian Sea.

5.3 State Practice in the Delimitation of the Caspian Sea Until 2018

Since the dissolution of the Soviet Union and the emergence of five Caspian Sea independent states, there have been multiple attempts to address the lack of explicit regulation of maritime borders within the Caspian Sea. The extension of land borders between Astara and Hosseingholi over the maritime areas was in 1935 officially recognized by the Soviet government as a state border in the Caspian Sea. The same was officially reconfirmed in 1954, and it was specified as a borderline in the airspace between the two countries in the common air navigation agreement of 1964. In 1970, the Soviet oil ministry divided its part of the Caspian Sea between the four Soviet republics (Azerbaijan, Kazakhstan, Russia, and Turkmenistan) with a

---

[26] In the case of Libya against Malta, the court has recognized the rights of third States to the extent that it has excluded from the zone of demarcation those parts to which a third-country raised claims, see: ICJ, Reports 1985, § 2, p. 16. Certain modifications with respect to this variant of the Tripoint model were expressed in the ICJ opinion Guinea against Guinea-Bissau. These regions were excluded from the demarcation, where other lines of delimitation were already established (ICJ, Reports 1985, § 93).

[27] ICJ, Rep. 1985, dissenting opinion of Schwebel, pp. 176–177.

middle line, where the Astara–Hosseingholi line was taken as the southern boundary of the Azerbaijani and Turkmen sectors.

Under the regime of the Soviet–Iranian agreements of 1921 and 1940, the two countries, being the exclusive coastal states, regarded the Caspian Sea as a closed sea standing fully under the sovereignty of the littoral states and remaining closed for access for other countries. However, after the collapse of the Soviet Union and the emergence of new sovereign states, the boundaries on the Caspian Sea became a subject of intergovernmental negotiations. As the negotiations were originally unsuccessful, it brought three bordering countries—Russia, Kazakhstan, and Azerbaijan—to a decision to carry out a unilateral division of the northern part of the Caspian Sea. In the period from 1998 to 2004, three bilateral treaties and one trilateral treaty, with additional protocols, were concluded that regulated the delimitation between the relevant national sectors of these countries as well as the regime of exploitation of natural resources in the northern part of the Caspian Sea. While referring to median line for seabed delimitation, the contracting states have sought for an equal sharing on the Caspian Sea.

Based on the Agreement between Kazakhstan and Russia from 1998, the Caspian seabed was divided into relevant sectors, and in accordance with the Additional Protocol of 2002, the water column remained for common use of both parties. Article 1 of the Treaty saw a division of the northern part of the seabed and subsoil of the Caspian Sea between the parties in accordance with the equity principle and the relevant parties' agreement. The prescribed middle line shall be equidistant from the baseline, considering existing islands and other geological features. The line was to be drawn from the coast line by 27 m, measured according to tide-gauge in Kronstadt. However, this method causes problems for present delimitations. First, the tide-gauge in Kronstadt was introduced in the eighteenth century, and thus it does not pay due regard to later changes of the sea level. Second, at the present time, cycles of several years are being used for sea-level measurement (for instance, a cycle of 19 years for the United States), which are not used at all for the tide-gauge in Kronstadt.

The 2001 Agreement between Azerbaijan and Kazakhstan and its Additional Protocol of 2003 divided relevant Caspian seabed sectors between the two parties along the middle line. The treaty disregarded the water column of the Caspian Sea. Each point on the middle line is located the same distance away from the nearest points on the coastline, including islands. The line was to be drawn from the coast line by 28 m, measured according to the tide-gauge in Kronstadt. The line specified in the treaty ran in the northwest of the Caspian Sea from the tripoint with Russia to the tripoint with Turkmenistan in the southeast. Further details regarding boundary intersection points and the points on the baseline were defined in the Additional Protocol.

The 2002 Agreement between Azerbaijan and Russia established two national sectors in the Caspian Sea, starting in the northwest on the mainland and continuing to the Azerbaijani–Kazakh–Russian tripoint. Article 1 of the Treaty requires national seabed and subsoil sectors in the northern part of the Caspian Sea to be delimited in accordance with generally recognized principles of international law in the form of a middle line, which suits the Caspian states and requires their consent. In addition, the article includes exact coordinates of the middle line, including the coordinates of the tripoint.

At last, an agreement between all three parties—Azerbaijan, Kazakhstan, and Russia—was signed in 2003, establishing a tripoint for the division of the northern part of the Caspian Sea among them.[28] According to Article 1, the point was located at the intersection of three Caspian seabed sectors at latitude $42°33'6''$N and longitude $49°53'3''$E. This so-called tripoint is a point at which the boundaries of three countries meet and which is equidistant from the nearest points of the coasts of the parties and the relevant third countries. This Agreement was to settle the extension of the rights of the contracting states in the northern Caspian Sea zones regarding the use of resources, till it would be replaced by a future treaty on the legal status of the Caspian Sea.[29] To delimit the entire seabed of the Caspian Sea, two additional tripoints would be required: in the central part—between Azerbaijan, Kazakhstan, and Turkmenistan—and in the south—between Azerbaijan, Iran, and Turkmenistan.

The delimitation process takes place in two stages. After the status of the seabed and its subsoil is first settled, the relevant provisions for the water column are to be specified. This approach was already applied in the North Caspian agreements, first between Azerbaijan and Russia in 2002 (Article 1) and then between Kazakhstan and Azerbaijan in 2003 (Article 1). The question of the legal status of the water column in the northern part of the Caspian Sea was not addressed in two of three North Caspian agreements because of the potential territorial claim of Azerbaijan. Only in the Agreement between Kazakhstan and Russia of 1998 (Article 1), after defining the status of the seabed between their coasts, was the water column left for common use.

All the three North Caspian agreements reassure that they do not prevent a comprehensive multilateral agreement among all the coastal states regarding the

---

[28]Kazakhstan on 4th December 2003, Azerbaijan on 9th December 2003.

[29]There are two different Tripoint methods, namely the so-called natural Tripoint, which is a an equidistant point between the states, which is located at the intersection of three equidistant lines; and an ad hoc Tripoint, located at any intersection located on the delimitation lines, see: Beazley (1993), pp. 256–259. This method was applied indirectly by the Permanent Court of Arbitration in 1999 in its judgment on the case of Yemen against Eritrea II (www.pca-cpa.org/ERYE2TOC.htm; Despeux 2000, pp. 459 et seq.). Without deciding on claims of third countries—especially of Saudi Arabia and Djibouti—the court set up the starting and the end points of the maritime delimitation between Yemen and Eritrea, which were equidistant from the coasts of these states and of third countries. Another method similar to Tripoint provides for delimitation, but for maritime zones to which third countries make a claim the delimitation has a temporary validity and thus it only has a potential character. In the case of Tunisia to Libya, in the area affected by the demands of third countries, the ICJ marked the direction of delimitation line using an arrow (ICJ, Rep. 1982, § 33 and 130, pp. 42 and 91). "The end point of the maritime boundary line which occurred in this way will match the future Tripoint between Tunisia, Libya, and Malta." This method differs from the strict "*res inter alios acta*" method, because a court is not authorized to delimit disputed areas.

legal status of the Caspian Sea.[30] Nevertheless, the legality of the North Caspian agreements was rejected by Iran, pointing out to its contradiction with existing Soviet–Iranian agreements.[31] The North Caspian Agreements are binding merely on their contracting parties and do not set any obligations for either Iran or Turkmenistan. According to international law provisions codified in the Vienna Convention on the Law of Treaties of 1969 (Article 34), a treaty does not create either obligations or rights for a third state without its consent. In practice, the North Caspian Agreements present a unilateral answer to the urgent economic need for the use of the Caspian resources. Their conclusion slows down the multilateral negotiations on the final status of the Caspian Sea conducted by all Caspian states but does not cancel them out. Even if they remain legally controversial among the Caspian littoral states, they seem to pave the way for a new pattern in dealing with the challenging goal of settling the legal status of the Caspian Sea. Without belittling the importance of multilateral cooperation for finding a mutually acceptable legal solution for the future status of the Caspian Sea, and in light of the pressing economic needs of the coastal states, the North Caspian Agreements may be seen as a contribution to the promotion of legal stability in the region.

On December 2, 2014, the Agreement between the Republic of Kazakhstan and Turkmenistan on the delimitation of the bottom of the Caspian Sea between the Republic of Kazakhstan and Turkmenistan was signed, and it entered into force on July 31, 2015. This Agreement provides for the sovereign rights of Kazakhstan and Turkmenistan over the established sectors of the seabed and subsoil of the Caspian Sea for the exploration, development, and exploitation of resources; laying of pipelines and cables; building of artificial islands, as well as berms, dams, overpasses, platforms, and other engineering constructions; and carrying out of other legitimate economic activities. In case the resources crossed the line of delimitation between the sectors of the contracting parties, the question of exploration and exploitation of the resources shall be defined in the separate agreement of these countries (Article 3). The water column was, however, not regulated in this Agreement. This Agreement shall not hinder the conclusion of the overall agreement on the legal status of the Caspian Sea by all of the coastal states and shall be interpreted as part of their legal relations.

The Caspian Sea Convention, adopted in 2018, has confirmed the binding force of the bilateral and trilateral agreements (Article 8, Para. 4), stating as follows:

> The exercise of sovereign rights of a coastal State under Para. 1 of this Article must not infringe upon the rights and freedoms of other Parties stipulated in this Convention or result in an undue interference with the enjoyment thereof (Article 8, Para. 4).

Its Article 20 further confirms:

---

[30] Article 9 of the Agreement between Kazakhstan and Russia, 1998; Article 5 of the Agreement between Azerbaijan and Russia, 2002, Article 5 of the Agreement between Azerbaijan and Kazakhstan of 2003.

[31] UN Doc. A/52/913 from 21st May 1998, Attachment; UN Doc. A/56/850 from 1st March 2002; UN Doc. A/56/1017 of 31st July 2002.

This Convention shall not affect rights and obligations of the Parties arising from other international treaties to which they are parties.

5.4 Delimitation of the Caspian Sea as Reflected in the Multilateral Negotiations Among the Caspian States Until 2018

None of the previously existing practice on the division of the Caspian Sea—neither the prolongation of the land boundary line between Astara and Hosseingholi remaining from the Soviet period nor the recently undertaken division of the northern part of the Caspian Sea—has offered a mutual agreement between the coastal states regarding state borders in the Caspian Sea. The need for a future clarification of its legal status, which was supposed to be based on the consent of all coastal states, has never been denied by them. The underlying requirement was, however, to balance countries' security concerns over adjacent maritime areas and, on the other hand, their economic interests linked to the use of Caspian resources.

Previously, in the draft of the Convention of the Caspian Sea status, it was proposed that the water column of the Caspian Sea split into territorial sea and fishery zone (Azerbaijan's proposal) or zones of national jurisdiction (Russia's proposal) and there be a water area that remains in common use by contracting parties and where free merchant navigation and freedom of fishery (Azerbaijan proposal), freedom of navigation and agreed fishery norms (Russia's proposal), and protection of the environment are guaranteed and that the water column of the Caspian Sea split into territorial sea, fishery zone, and the high sea (Kazakhstan's and Turkmenistan's proposal) and the water column of the Caspian Sea consist of national zones, where freedom of navigation, agreed fishery norms, and protection of the environment are guaranteed (Iran's proposal). The seabed and its subsoil are to be delimited for the purpose of exercising rights on the exploitation of resources and other lawful economic activities related to the exploitation of the resources of the seabed and its subsoil (rejected by Iran).

In the framework of interstate negotiations, as reflected in the draft of the Caspian Status Convention, the internal waters were defined as "waters on the landward side of the baseline" (Article 1(14)). However, because there was no uniform coastal state standing on the recognition of the territorial sea zone in the Caspian Sea, there were no exact provisions concerning the internal waters. If the negotiating parties would follow the United Nations Convention on the Law of the Sea (UNCLOS) and accept the concept of the Territorial Sea included there, the definition of internal waters of the Caspian Sea could have been respectively derived also from UNCLOS provisions.

In the framework of the negotiations on the Caspian Sea's status, Russia, for many years, had proposed for the establishment of a zone of national jurisdiction and rejected the idea of introducing a territorial sea zone. According to the law of the sea, a zone of national jurisdiction means a maritime area that consists of internal waters and the territorial sea.[32] Thus, within such a zone, a coastal state exercises unrestricted sovereignty. One can see the exclusive economic zone (EEZ) as part of this zone, in which a coastal state exercises economic sovereignty. However, the definition of the zone of national jurisdiction proposed by Russia is different: "a water body adjacent to the coast and extending no more than 15 sea miles from the baseline, where freedom of navigation and agreed fishery norms as well as environmental protection are secured" (Article 5(6).1). Russia's concept of "zone of national jurisdiction" did not recognize any sovereignty of coastal states there, explaining that "within zones of national jurisdiction coastal states exercise the control necessary to prevent infringement of its customs, fiscal, health and veterinary laws as well as enjoy exclusive fishery rights" (Article 7(8)). The Russian definition of zone of national jurisdiction was a direct reference to the UNCLOS provisions on the so-called contiguous zone and exclusive fishing zone. In either of the two zones, a coastal state exercises no full territorial sovereignty but only so-called sovereign rights, which are limited to certain police matters. According to UNCLOS, the contiguous zone may not extend beyond 24 nautical miles from the baselines, where a coastal state may exercise the control necessary to avoid and punish infringements of its customs, fiscal, immigration, or sanitary laws committed within its territory or territorial sea.[33] However, the nature and territorial scope of a coastal state's control within the contiguous zone remained unclear.[34] Finally, the idea of zone of national jurisdiction was not adopted, but the zone of territorial waters was established in the Caspian Sea Convention, guaranteeing state sovereignty over this area.

During the negotiations, Caspian countries came up with diverging proposals on how to define waters remaining outside the Fishing Zone or so-called Zone of National Jurisdiction. These ideas were reflected in the Draft of the Caspian Sea Convention as (1) "...waters being subject to joint use by the states where free merchant shipping and freedom of fishing [proposed by Azerbaijan], or in which the freedom of navigation and coordinated fisheries standards [proposed by Russia] and the environment are protected."; (2) "...the High Seas [proposed by Kazakhstan and Turkmenistan], which belong neither to Territorial Sea (National zone), nor to exclusive Fishery Zone nor to Internal Waters" [proposed by Turkmenistan]; (3) "...national zones, with ensured freedom of navigation, the agreed fisheries standards, and the protection of environment" [proposed by Iran].

The original draft of the Caspian Sea Convention did not provide for a special regime for the maritime territories outside of the Fishing Zone or Zone of National Jurisdiction. The Draft Convention defined such territories following the concept of the high seas known in the law of the sea, where high sea is based on the freedom and cannot to be subject to the territorial claims of coastal states. The concept

---

[32] See: (Dupuy and Vignes 1991), p. 291.
[33] Article 33(2, 3).
[34] See: (Wooldridge 1992), pp. 781 et seq.

guaranteeing freedoms of navigation and fishing proposed by Azerbaijan reflected international law's rules on the high seas, although it did not provide for other recognized freedoms. The proposal regarding the high sea zone in the Caspian Sea made by Kazakhstan and Turkmenistan did not fully reflect the relevant law of the sea standards. All coastal countries agreed that according to previously binding rules, navigation on the Caspian Sea shall remain free and unrestricted to all the neighboring countries. The principle of freedom of navigation and ensuring safety covered merchant ships flying the flag of a Caspian coastal state but excluded any foreign ships from the principle as it was the practice before the dissolution of the Soviet Union.[35] The introduction of such an understanding concerning the freedom of navigation outside of the fishing zone or the zone of national jurisdiction into the Draft Caspian Status Convention was meant as a significant limitation to the traditional rights on fishery.

In the course of the interstate negotiations on the legal status of the Caspian Sea, the following ways of delimitation were proposed in the Draft Caspian Sea Convention: the delimitation of the seabed and its subsoil in sectors/zones of national jurisdiction should have been based on the median line (Iran against) and with the consent of all states whose coasts are opposite or adjacent to each other, with due regard to the norms of international law and state practice established in the Caspian Sea (Turkmenistan against) and according to equitability (Azerbaijan against) (Article 8(9).1). Kazakhstan agreed with such formulation but added a proposal: the delimitation of the Caspian seabed and its subsoil shall be conducted upon an agreement by the states whose coasts are opposite or adjacent to each other (Azerbaijan against). Russia agreed with both versions of Article 8(9), proposing to add a provision directly referring to the North Caspian Agreements: in case the parties have already signed a relevant agreement concerning the delimitation of the seabed and its subsoil, all questions relating to delimitation shall be decided in accordance with such agreement (Iran against).

The procedure laid down in the draft of the Convention was similar to the UNCLOS provisions regarding delimitation based on the equidistance method. However, the draft remained inconsistent with other provisions of UNCLOS and thus created serious ambiguity. For example, according to UNCLOS, a coastal sea remains under territorial sovereignty. However, the Caspian littoral states were far from unanimously accepting such territorial sea concept and that this concept is at all applicable to the Caspian Sea. In the draft, the principle of the median line was mentioned in relation to the delimitation of the sectors of the seabed and subsoil and not, as is the case in UNCLOS, in relation to the entire territorial sea. The conceptual differences between the draft and the globally accepted understanding of the territorial sea concept would have resulted in serious errors in the future status of the Caspian Sea. Azerbaijan's disagreement with the application of the equity principle in the implementation of the median line was rooted in its fear to revive the Iranian demand of dividing the Caspian Sea into five equal parts. This would show,

---

[35] Article XIV, XV of 1935 Agreement; Article 12, 13 of 1940 Agreement.

however, a false understanding of this principle by Azerbaijan. In the negotiations leading up to the Draft Caspian Status Convention, Azerbaijan, Kazakhstan, and Russia called for the acceptance of the existence and recognition of the binding force of previously signed agreements on the delimitation of the northern part of the Caspian Sea, allowing for the exercise of sovereign rights that enable parties to use the natural resources of its seabed. Parties to the North Caspian Agreements agreed that these treaties will not prevent the conclusion of an overall agreement among the Caspian littoral states regarding the legal status of the Caspian Sea.

However, in the Draft Caspian Status Convention, Azerbaijan, Kazakhstan, and Russia requested that all arising questions shall be decided in accordance with existing contracts related to Caspian delimitation, which included a preliminary solution to some aspects of the Caspian Sea's status. This proposal was met with opposition from Iran; Turkmenistan, on the other hand, abstained.

5.5 Territorial Division of the Caspian Sea According to Caspian Sea Convention

The intergovernmental negotiations initiated just after the collapse of the USSR to develop a new legal status for the Caspian Sea resulted in the development of the Caspian Sea Convention in 2018, as elaborated further in this chapter. It was accompanied by some statements reflecting on the position of the Caspian states offered during the intergovernmental negotiations conducted during the last two decades. They discussed a possible introduction of maritime zones in the Caspian Sea, in accordance with UNCLOS, to determine the scope of coastal states' sovereignty—whether full sovereignty or merely sovereign rights—and thus the extent of their rights to the use of the natural resources. Throughout the years, the Caspian states differed extensively as to the nature of the proposed zones, until 2018, when they reached a mutual understanding.

In international law, a clear border setting reduces the risk of extending a state's rights to maritime zones. The relevant law applicable in this instance is the Law of the Sea. However, though the Caspian Sea Convention steered clear of any reference to the Law of the Sea as a legal model for the Caspian Sea regulations, it follows UNCLOS, introducing equivalently named categories of maritime zones, though their notions differ from those provided in UNCLOS. UNCLOS, which sets the standard for the territorial extension of maritime zones, clarifies the extent of state sovereignty in such zones, whether full sovereignty or state sovereign rights (so-called functional rights).

According to UNCLOS, all maritime zones are defined using the distance criterion: territorial sea of 12 nautical miles (Article 3), the contiguous zone of 24 nautical miles (Article 33(2)), the exclusive economic zone of 200 nautical miles (Article 57), the continental shelf of a maximum of 350 nautical miles or 100 nautical miles from the 2500 m isobath (Article 76(5) and (6)), and the high seas (Article 86). These

5.5 Territorial Division of the Caspian Sea According to Caspian Sea Convention

zones provide for separate regulations of regimes like shipping, use of natural resources, laying pipelines, etc.

Article 5 of the Caspian Sea Convention, which will be discussed below in details, provides for the following:

> The water area of the Caspian Sea shall be divided into Internal Waters, Territorial Waters, Fishery Zones and the common Maritime Space (Article 5).

> The sovereignty of each Party shall extend beyond its land territory and Internal Waters to the adjacent sea belt called Territorial Waters, as well as to the seabed and subsoil thereof, and the airspace over it (Article 6).

Before discussing the maritime zones of the Caspian Sea, as finally defined in the Caspian Sea Convention of 2018, it has to be stated that zones are all measured from baselines determined in accordance with the provisions of the law of the sea. A baseline allows to mathematically estimate the course of a middle line (often used for the delimitation of the territorial sea zone), every point of which is equidistant from the nearest points on the baseline. In general, this term means a normal baseline for measuring the breadth of the zones, which is the low-water line along the coast, as marked on large-scale charts officially recognized by the coastal state.[36] A baseline is also provided for islands,[37] reefs,[38] low-tide elevations,[39] and rocks.[40] Coastal state rights can be also established by applying the principle of straight baselines, which nowadays seems to become a part of customary international law. The system of straight baselines is being adopted by most coastal states.[41] The International Court of Justice has already referred to it in a dispute over the Norwegian coastal waters.[42] The drawing of straight baselines must not depart, to any appreciable extent, from the general direction of the coast, and the sea areas lying within the lines must be sufficiently closely linked to the land domain to be subject to the regime of internal waters.[43] The method of straight baselines joining appropriate points may be employed in, first, localities where the coastline is deeply indented and cut into or if there is a fringe of islands along the coast in its immediate vicinity[44] and, second, mouths of rivers,[45] bays,[46] and archipelagos.[47]

---

[36] Article 5 UNCLOS.

[37] Article 10 (2): 1958 Convention on the Territorial Sea and the Contiguous Zone 1958; Article 121 (2) UNCLOS.

[38] Article 6 UNCLOS.

[39] Article 8: 1958 Convention on the Territorial Sea and the Contiguous Zone 1958: Article 13: UNCLOS.

[40] Article 121 (3): UNCLOS.

[41] See: Bernhardt (1984), p. 218.

[42] ICJ Reports 1951, pp. 116 et seq.

[43] Article 7 (3).

[44] Article 7 UNCLOS.

[45] Article 9 UNCLOS.

[46] Article 10 UNCLOS.

[47] Article 47, UNCLOS.

The Caspian Sea Convention provides (Article 7) that each country sets the breadth of its territorial sea measured from a baseline. This provision does not indicate either the normal or the straight baseline as a method of measurement. The Caspian Sea Convention provides for a uniform definition of the baseline notion as follows:

Article 1:

"Baseline" – the line consisting of normal and straight baselines.

The Caspian Sea Convention envisages both normal and straight baselines and defines them as follows:

"Normal baseline" – the line of the multi-year mean level of the Caspian Sea measured at minus 28.0 meters mark of the 1977 Baltic Sea Level Datum from the zero-point of the Kronstadt sea-gauge, running through the continental or insular part of the territory of a Caspian littoral State as marked on large-scale charts officially recognized by that State.

"Straight baselines" – straight lines joining relevant/appropriate points on the coastline and forming the baseline in locations where the coastline is indented or where there is a fringe of islands along the coast in its immediate vicinity. The methodology for establishing straight baselines shall be determined in a separate agreement among all the Parties.

The definitions included in the Caspian Sea Convention are very similar to the provisions of UNCLOS in relation to the regulation of the baseline, wherein the low-water line is to be regarded as a starting point for determining the breadth of the territorial sea and the subsequent sea zones. The methodology for establishing straight baselines was not settled by the Caspian Sea Convention but was left for the determination of the parties in separate agreements.

5.5.1 Internal Waters

The waters on the landward side of the baseline of the territorial sea form part of the internal waters of the state.[48] In this area, the coastal state exercises total and absolute sovereignty, equal to the sovereignty over the land territory, air space, and internal waters, as well as its seabed and subsoil.[49] The legal status of the territorial sea extends to the internal waters; thus, the comprehensive and undisturbed exercise of sovereign rights of the coastal state within internal waters is warranted.

Such a definition reflects the regulation of UNCLOS on internal waters. According to UNCLOS, internal waters are waters on the landward side of the baseline of the territorial sea (Article 8), where the coastal state exercises unlimited sovereignty over the water column and its seabed and subsoil.

---

[48] Article 8 UNCLOS.
[49] ICJ Case: Nicaragua, ICJ Report 1986, 111.

5.5.2 Territorial Waters in the Caspian Sea Convention

According to international law, the sovereignty of a coastal state extends beyond its land territory and internal waters and, in the case of an archipelagic state, its archipelagic waters, to an adjacent belt of sea, described as the territorial sea.[50] This sovereignty extends to the air space over the territorial sea, as well as over its bed and subsoil.[51]

The application of the territorial sea concept to the Caspian Sea used to be a major point of contention during the negotiations over its future status. In the territorial sea, coastal states exercise full and unrestricted sovereignty over all activities, including fishing, mining, environmental protection, customs, etc. The rights of the coastal state in the territorial sea are restricted only with respect to shipping. Ships of all states, whether coastal or landlocked, enjoy the right of innocent passage through the territorial sea. In the coastal areas, where most shipping routes are located, coastal state rights are limited by the principle of innocent passage,[52] which is already firmly established in states' practice worldwide.[53] The question of seaward extension of the territorial sea has always been particularly controversial in the history of the law of the sea. It is regarded as one of the greatest achievements of UNCLOS as it included, for the first time ever, a firm legal framework of the outer limits of the territorial sea.[54] UNCLOS states that every state has the right to establish the breadth of its territorial sea up to a limit not exceeding 12 nautical miles, measured from baselines.[55]

Following the standards set in UNCLOS, the Caspian Sea Convention does not expand the freedom of navigation to the territorial waters (Article 10, Para. 1), envisaging almost unlimited control of coastal states in the territorial sea zone. The introduction of a territorial sea zone in the Caspian Sea has resulted in an impossible interference by a coastal state in the internal affairs of another coastal state involving its territorial sea. It has recognized state boundaries in the Caspian Sea, which historically were never legally settled but only claimed by Azerbaijan, Kazakhstan, and Turkmenistan. The territorial water zone is defined as a belt of sea to which the sovereignty of a coastal state extends (Article 1), covering the water

---

[50] Article 17–32 UNCLOS.

[51] Article 2 UNCLOS.

[52] Passage related to Article 18 UNCLOS means navigation through the Territorial Sea for: traversing that sea without entering Internal Waters or calling at a roadstead or port facility outside Internal Waters; or proceeding to or from Internal Waters or a call at such roadstead or port facility. Passage shall be continuous and expeditious. However, passage includes stopping and anchoring, but only in so far as the same are incidental to ordinary navigation or are rendered necessary by *force majeure* or distress or for rendering assistance to persons, ships or aircraft in danger or distress. The passage is innocent so long as it is not prejudicial to the peace, good order or security of the coastal State (Article 19 UNCLOS).

[53] Wolfrum (1990), pp. 20–23.

[54] See: (Wolfrum 1990), pp. 20–23.

[55] Article 3 UNCLOS.

column, as well as the seabed and subsoil thereof (Article 6). The breadth of the territorial waters shall not exceed 15 nautical miles, and the outer limit of the territorial waters is considered to border a coastal state (Article 7, Para. 2). The delimitation of internal and territorial waters between states with adjacent coasts shall be affected by an agreement between those states (Article 7, Para. 3). Most shipping routes are located along coastal areas, where riparian states exercise their sovereignty. To allow for international navigation and trade, the law of the sea limits the authority of the coastal states in the territorial zones (Article 58, Para. 1 UNCLOS). Also, Article 11, Para. 1, of the Caspian Sea Convention allows "vessels, flying the flags of the other coastal states, for navigation through Territorial Waters with a view to traversing those waters without entering Internal Waters or calling at a roadstead or port facility outside Internal Waters or proceeds to or from Internal Waters." Ships enjoy the right of innocent passage through the territorial seas of the riparian states (Article 11, Caspian Sea Convention). The terms and conditions of innocent passage within the territorial waters of the Caspian Sea are comparable international law standards (Article 18, UNCLOS). This right, however, has been reserved exclusively for the contracting Caspian states and extend to other countries as it is a standard in the UNCLOS regulations over the Territorial Seas.

5.5.3 The Fishery Zone and the Common Maritime Space

The legal status of the Caspian maritime zones, specifically the legal status of both the water column and the sea bottom below the water column seaward of the maritime boundaries, carries different names and differs in regulation from the zones foreseen in the Law of the Sea. The legal status of the water column is relevant for the regime of transport of natural resources by vessels within and outside of the Caspian Sea. The legal status of the bottom of the Caspian Sea below the water column defines the development of the natural resources of the seabed and the subsoil. It will be reviewed in the following subchapters, separately from the status of the water column, because their regulations significantly differ from each other. In terms of the legal status of the water column in the Caspian Sea, there are two identifiable zones: the fishery zone and the common maritime space.

The fishery zone (up to 10 nautical miles wide), envisaged in the Caspian Sea Convention, is defined as a belt of sea where a coastal state holds an exclusive right to harvest aquatic resources (Article 1). The water within this zone is free for the shipment of goods by all Caspian Sea states and, therefore, important for the development of the commercial transport of fossil fuels, and other natural resources, from the Caspian Sea. Coastal states, however, reserve the right in the fishery zone to ensure compliance by foreign vessels with their laws and regulations, including boarding, inspection, hot pursuit, detention, arrest, and judicial proceedings, in order to exercise exclusive rights to harvest aquatic biological resources as well as for the purposes of conserving and managing such resources (Article 12, Para. 3):

5.5 Territorial Division of the Caspian Sea According to Caspian Sea Convention

Party, in the exercise of its sovereign right to the subsoil exploitation [...] of resources of the seabed and subsoil, may take measures in respect of ships of other Parties (Article 12, Para. 3).

This provision does not pose a difference from the Law of the Sea standards. UNCLOS, however, does not envision a fishery zone as it provides for the so-called EEZ, which in terms of legal regime of fishing can be, by analogy, applied to the fishery zone. The EEZ is an area beyond and adjacent to the territorial sea (Article 57, UNCLOS), where a coastal state has a sovereign right for the purpose of exploring, exploiting, conserving, and managing the natural resources (Articles 55 and 56, UNCLOS), whether living or nonliving. Freedom of the sea, including navigation and the laying of submarine pipelines, applies accordingly (Article 58, UNCLOS). In the exercise of its sovereign rights to explore, exploit, conserve, and manage the living resources in the exclusive economic zone, the coastal state may take such measures, including boarding, inspection, arrest, and the conduct of judicial proceedings (Article 73, Para. 1). The main difference between the legal status of the EEZ and that of the fishery zone is that in the fishery zone, the sovereign right of the coastal state is limited to living resources.

According to the Caspian Sea Convention, a water area located outside of the outer limits of a fishery zone is called the common maritime space and is open to any type of uses by all the parties to the Convention (Article 1). This zone provides for the joint sovereignty of all riparian states over the surface waters (Article 5).[56] This regime is rather similar to the regime of the surface waters of international lakes, which, unlike those of seas, can be used by states bordering them.[57] The status of this zone is, to some extent, similar to UNCLOS provisions for the so-called high sea zone in terms of regulation over its water column status. The high sea includes all parts of the sea that are placed beyond the limits of national jurisdiction and are not included in the EEZ, territorial sea, or internal waters of a state (Article 33, UNCLOS), and the main legal principle characterizing the high sea is "freedom of the sea." This also includes regulation for its navigation, as well as the laying of submarine cables and pipelines therein. The Caspian Sea Convention opens the common maritime space to all types of use, including the shipment of natural resources to all coastal states (freedom of the sea). The access to this zone is, however, limited merely to the Caspian Sea coastal states, which has an impact on international trade because it excludes vessels of other countries from accessing this zone, and in this way significantly differs from the regulation of the high sea as provided by UNCLOS. Another difference between the high sea regime and that of the common maritime space is that in the latter, "each Party, in the exercise of its sovereign right to the subsoil exploitation [...] of resources of the seabed and subsoil, may take measures in respect of ships of other Parties" (Article 12, Para. 3).

The bottom, and the subsoil, of the fishery zone and common maritime space is regulated by the legal regime for the development and use of natural resources, as

---

[56]Ibid.
[57]Oxman (1996).

well as the regime of maritime pipelines. According to the Caspian Sea Convention, it shall be shared into sectors. "Sector" is defined by the Caspian Sea Convention as part of the seabed and subsoil delimited between the parties for the purposes of subsoil exploitation and other legitimate economic activities in this territory (Article 8, Para. 1). The delimitation of the Caspian Sea seabed, and the dividing of the subsoils into sectors, aims at enabling the coastal states to exercise their sovereign rights to develop subsoil resources within the sectors. Such a policy reflects the legal regime of the continental shelf, recognized in UNCLOS, regarding the development of resources and the laying of submarine pipelines. According to UNCLOS, the continental shelf is an area of a maximum of 300 nautical miles or 100 nautical miles from the 2500-meter isobath (Article 76, Para. 2), which comprises the seabed and subsoil of the submarine areas that extend beyond its territorial sea and where a coastal state exercises exclusive sovereign rights for the purposes of exploring and exploiting the natural resources (Articles 76 and 77, UNCLOS).

The legal regime introduced by the Caspian Sea Convention significantly differs from the legal concept developed in UNCLOS for the bottom of the maritime zones, which are located outside of the outer limits of the national boundaries of the coastal states and are covered by adjacent waters under common use (high sea). The bottom of the sea and the subsoil under the high sea, the so-called Area, remain a common heritage of humanity, free from sovereign rights of coastal states. The Caspian Sea Convention provides for a significantly different approach to the territories comparable to the so-called Area as provided in UNCLOS. The Caspian Sea Convention divides the bottom and the subsoil of the Common Maritime Space into sectors, where the coastal states exercise their sovereign rights to develop natural resources (Article 6).

The Caspian Sea Convention, when defining the principle of delimitation of the Caspian seabed and subsoil in zones, does not refer to the status of the water column. Such an approach takes its roots in the provisions of UNCLOS concerning the delimitation of the exclusive economic zone. Its peculiarity is that in contrast with the delimitation of the continental shelf, the exclusive economic zone is composed of the water column and the seabed and therefore its delimitation may be conducted in separate processes for the water column and for the seabed. The course of delimitation of the water column may initially vary from the course of delimitation of the seabed.[58] The appropriateness of this distinction was confirmed by the ICJ, which pointed out to the need to consider the peculiarities of living and nonliving resources.[59] A similar regulation was introduced in the 1958 Geneva Convention on the continental shelf, where waters above the continental shelf are to be considered as high seas.[60] A similar process of delimitation conducted in two separate stages was applied in the North Caspian Agreements between Azerbaijan and Russia in 2002 (Article 1) and between Azerbaijan and Kazakhstan in 2003 (Article 1).

---

[58]See: (Ipsen 2004), p. 869.
[59]ICJ, Rep. 1984, p. 246.
[60]See: (Oda 1995), p. 306.

5.5.4 Methods of Maritime Delimitation of the Caspian Sea

Having presented the necessary conditions for a possible conducting of maritime delimitation, it is now time to discuss the existing methods of delimitation, which affect, among other things, the utilization of Caspian resources. One of the rules mainly used in international law is the so-called principle of equidistance.

The application of this principle points out to the method of median line, every point of which is equidistant from the nearest points on the baselines, from which the breadth of the territorial seas of neighboring states is measured.[61] The following variants of the equidistant line are known in states' practice: simplified,[62] adapted,[63] and strict.[64] The legal sources for the rule on delimitation and thus the equidistant principle are the Geneva Conventions of 1958 and UNCLOS, as well as international case law. The Geneva Conventions provide for a free choice of delimitation methods and do not prescribe equidistance as a compulsory method. Only if there is no other agreement between them is the equidistance method, considering reasons of "special circumstances," preferred. While conducting delimitation of the continental shelf or exclusive economic zone an equitable solution must be reached,[65] what, however, is not equal with the application of the ex aequo et bono principle.[66] The equidistance method shall not be seen as a norm of customary law since the necessary *opinio juris* is failing; however, this method has already found indisputable roots in the international legal practice of states.[67]

---

[61] Article 12, § 1 and Article 24 § 3 of the Convention on the Territorial Sea and the Contiguous Zone 1958; Article 6, § 1 and 2 of 1958 Convention on the Continental Shelf; Article 15: UNCLOS.

[62] Denmark/United Kingdom, see (Elferink 1999), p. 548.

[63] For example Denmark/Iceland, see (Elferink 1998), pp. 607 et seq.

[64] United Kingdom/United States of America, In: IJMCL, vol. 9, No. 2, 1994, pp. 258–259.

[65] See ICJ judgement in the case of the North Sea continental shelf of 1969 (ICJ, Reports 1969, pp. 3 et seq.). The principle of equity has come to apply for the first time in the North Sea Continental Shelf case, where the ICJ concluded that in addition to consideration of the relevant circumstances in the region, the application of the "equity principles" is a binding rule of customary law in case of delimitation between states, who are not party to contractual law (ICJ, Reports 1969, § 101 c 1, p. 54). With this decision, the ICJ has caused a long-term disagreement of the states with respect to the nature of the application of the equidistance method for the delimitation of maritime zones beyond the Territorial Sea. This conflict of contractual and judicial law has not been repealed by the provisions of UNCLOS. The UNCLOS requires an equitable solution in the continental shelf or EEZ delimitation in the absence of agreement among the involved states, which can be achieved in the application both of equidistance and equity methods (Article 74, 83 UNCLOS). The final set of rules that explain the current legal application of the equity principle have been established by the judgment of Qatar against Bahrain. This confirmed that application of the equidistant principle is the first and the application of the equity principle is the second stage of each delimitation case.

[66] ICJ, 1969, § 88, 49.

[67] Gornig and Despeux (2002), pp. 184 et seq. A certain degree of inconsistency is shown in ICJ judgments in the cases Libya against Malta from 1985 (Libya/Malta, ICJ, Reports 1985, pp. 12 ff. and Denmark against Norway from 1993 (ICJ, Reports 1993, pp. 37 et seq.). The ICJ has used the equidistance method as basis, which must be examined in the second phase based on special or relevant circumstances and the principle of equity. This unification was first referred only to

Regarding the method of delimitation of the Caspian Sea, the Convention provides for the following:

> The outer limit of the Territorial Waters shall be the line every point of which is located at a distance from the nearest point of the baseline equal to the breadth of the Territorial Waters.
>
> For the purpose of determining the outer limit of the Territorial Waters, the outermost permanent harbour works which form an integral part of the harbour system shall be regarded as forming part of the coast. Off-shore installations and artificial islands shall not be considered as permanent harbour works.
>
> The outer limit of the Territorial Waters shall be the state border (Article 7, Para. 2).

The Caspian Sea Convention envisages the equidistance principle and sets some special conditions for its application to the delimitation of the territorial waters of the Caspian Sea, including the validity of signed delimitation agreements and compliance with the norms of international law:

> Delimitation of internal and Territorial Waters between States with adjacent coasts shall be effected by agreement between those States with due regard to the principles and norms of international law (Article 7, Para. 3).

In terms of the delimitation of the Caspian Sea and its subsoil into sectors, the following was declared by the Caspian Sea Convention in its Article 8, Para. 1:

> Delimitation of the Caspian Sea seabed and subsoil into sectors shall be effected by agreement between States with adjacent and opposite coasts, with due regard to the generally recognized principles and norms of international law, to enable those States to exercise their sovereign rights to the subsoil exploitation and other legitimate economic activities related to the development of resources of the seabed and subsoil.

It has included the obligation to take into account previous state practice in the Caspian Sea and the validity of previous relevant delimitation agreements:

> The exercise of sovereign rights of a coastal State under Para. 1 of this Article must not infringe upon the rights and freedoms of other Parties stipulated in this Convention or result in an undue interference with the enjoyment thereof (Article 8.4).

The division of the seabed, and subsoil, of the Caspian Sea under the fishery zone and the common maritime space into national sectors follows the pattern in the North Caspian Agreements between Kazakhstan, Azerbaijan, and Russia. The sectoral approach to the division of the seabed and subsoil, adopted in the Caspian Sea Convention, may indicate that the division of the bottom of the Caspian Sea, undertaken previously in the northern Caspian Sea, is now to be officially recognized and accepted by all riparian states. Article 4 of the Caspian Sea Convention states that the parties shall conduct their activities in the Caspian Sea, including the exploration and exploitation of the seabed and subsoil resources, in accordance with "other agreements" between the parties. Article 8, Para. 1, states clearly that the Caspian Sea seabed and subsoil are to be delimited into sectors, which shall be

---

opposite states (ICJ 1993, § 50, p. 60) and then also to adjacent countries (Qatar/Bahrain, ICJ 2001, in conjunction with § 170 & 224).

affected by an "agreement" between states with adjacent and opposite coasts. Further, Article 8, Para. 4 states that the "exercise of sovereign rights of a Coastal State under Para. 1 of this Article must not infringe upon the rights and freedoms of other Parties stipulated in this Convention or result in an undue interference with the enjoyment thereof." Also, Article 20 says that "This Convention shall not affect rights and obligations of the Parties arising from other international treaties to which they are parties." One may assume that the future delimitation of the sectors of Turkmenistan and Iran will similarly follow the pattern of the middle line, and the sectors assigned within the North Caspian Agreements to Azerbaijan, Kazakhstan, and Russian will remain unchanged.

Russia agreed with both versions of Article 8(9).1, proposing to add a provision directly referring to the North Caspian Agreements: "In case where parties have already signed a relevant agreement concerning the delimitation of the seabed and its subsoil, all questions relating to the delimitation to be decided in accordance with such agreements" (Iran against).

In terms of submarines and other underwater vehicles, the Caspian Sea Convention states that *submarines and other underwater vehicles of one Party shall be required to navigate on the surface and show their flag when passing through the Territorial Waters of another Party* (Article 11, Para. 5).

5.6 Conclusion

The setting of state maritime borders is crucial for the sustainable development of a country and for its peaceful and fruitful cooperation with its neighboring states. The delimitation of state borders requires a clear international legal framework based on the political consent of all states sharing a common water pool. Borderlines define the scope of state sovereignty, settling each state's rights to use the natural resources of the water area, lay pipelines, conduct shipping activities, etc. According to the customary law of the sea, every coastal state has the right to establish and use maritime zones, which shall be delimited according to common international standards.

Previously, there was no clear legal framework for state borders in the Caspian Sea. Neither agreements nor the legal praxis between the Soviet Union and Iran managed to define the legal status of the Caspian Sea in a way that would be recognized as binding by the coastal states, which emerged after the dissolution of the Soviet Union. It resulted in a long-term dispute around the status of the Caspian Sea, which lasted until 2018, and the conclusion of the Caspian Sea Convention. The main object of the dispute was possible application of the legal concepts of a sea, lake or condominium for the Caspian Sea while defining the Status of the Caspian Sea. Additional aspects within the negotiation process on the delimitation of the Caspian Sea, which came up in the late 1990s, took two different forms: first, continuing multilateral negotiations on future Caspian Sea delimitation, presented in the Draft Caspian Sea Status Convention, and, second, conclusion of bi- and trilateral agreements on sharing northern parts of the Caspian seabed for using resources located there. The first one would have defined maritime zones in the Caspian Sea; however, there were serious disagreements as to their scope. The greatest challenge was related especially to the introduction of a concept of a coastal sea zone, which would recognize states' sovereignty over such a zone and excludes most rights of other states in this area. This debate was also linked to the question of introducing a zone excluded from the coastal states' sovereignty, guaranteeing states' maritime freedoms.

In 2018, the Caspian Sea Convention was finally adopted, which introduced a set of clear principles for the territorial delimitation of the Caspian Sea. The Caspian Sea Convention has provided the framework for defining the scope of the territorial sovereignty of the Caspian Sea states. The application of the middle line and straight baselines for the delimitation of the Caspian Sea, proposed in the Caspian Sea Convention, has resulted in the establishment of national sectors, without referring anymore to the transboundary lake or sea concepts. As the Caspian Sea is 200 nm wide on average, the application of the proposed delimitation method results in establishing national sectors of ca. 20.6% for Azerbaijan, 14.6% for Iran, 30% for Kazakhstan, 15.6% for Russia, and 19.2% for Turkmenistan. However the Caspian Sea Convention steered clear from referring to the UNCLOS for regulation of the Caspian Sea status, it has followed UNCLOS provision widely. The Caspian Sea Convention divided the Caspian Sea water into the following zones: internal waters, territorial waters, fishery zones, and common maritime space.

Adjacent to the territorial waters, each coastal states will establish a 10-nautical-mile-wide fishery zone, similar to the concept under international law, where UNCLOS (Article 73, Para. 1, and Article 58) provides for a so-called exclusive economic zone and limits coastal states' sovereign rights to living resources in this zone. The Caspian Sea Convention reserves the right to ensure compliance by foreign vessels with national laws and regulations, including on boarding, inspection, hot pursuit, detention, arrest, and the conduct of judicial proceedings in order to exercise exclusive rights to harvest aquatic biological resources and for the purposes of conserving and managing such resources (Article 12, Para. 3). Outside of the outer limits of the fishery zone, the common maritime space was established, similar to the UNCLOS regulation on the so-called high sea, guaranteeing international freedom of the sea, but which is limited merely to the Caspian Sea coastal states. The bottom and subsoil of the fishery zone and the common maritime space will be divided into so-called sectors, where the coastal states can exercise their sovereign rights to develop resources. Such regulation differs from UNCLOS and Law of the Sea standards for the bottom of maritime zones located under the high sea. This so-called Area at the bottom of the high sea remains a common heritage of humanity. The delimitation of the seabed and subsoil sectors in the Caspian Sea shall be affected by agreements between the coastal states, taking into account international law as well as previously adopted agreements on the Caspian Sea. It provides for a clear reference to the so-called North Caspian Agreements between Kazakhstan, Azerbaijan, and Russia, and later Turkmenistan, which have divided the seabed and subsoil of the northern part of the region, allowing for the use of natural resources.

References

Beazley, P., 1993. Technical considerations in maritime boundary delimitation. In: Charney JI, Alexander LM (eds) International maritime boundaries. Martinus Nijhoff, Dordrecht

Bernhardt R (1984) Einfluß der UN-Seerechtskonvention auf das geltende und zukünftige internationale Seerecht. In: Delbrück J (ed) Das neue Seerecht. Duncker & Humblot, Berlin

Boggs S (1951) Delimitation of seaward areas under national jurisdiction. Am J Int Law 45 (2):240–266

Charney J, Alexander L (eds) (2003) International maritime boundaries. Martinus Nijhoff, Dordrecht

Dupuy RJ, Vignes D (eds) (1991) A handbook of the new law of the sea, 1st edn. Martinus Nijhoff, Dordrecht

Elferink AO (1998) Denmark/Iceland/Norway-bilateral agreements on the delimitation of the continental shelf and Fishery Zones, 13th edn. IJMCL

Elferink AO (1999) United Kingdom and Denmark/Faroe Islands maritime delimitation between the United Kingdom and Denmark/Faroe Islands, 14. IJMCL, (ред. б.м.)

Gornig G, Despeux G (2002) Seeabgrenzungsrecht in Der Ostsee: Eine Darstellung Des Voelkerrechtlichen Seeabgrenzungsrechts Unter Besonderer Berucksichtigung Der Praxis Der Ostseestaaten. Peter Lang AG, Bern, Switzerland

Ipsen K (2004) Völkerrecht, 5th edn. CH Beck, München

Oda S (1995) Exclusive economic zone. EPIL (б.м)

Orazgaliyev S, Araral E (2019) Conflict and cooperation in global commons: theory and evidence from the Caspian Sea. Int J Commons 13(2):962–976

Oxman B (1993) Political, strategic and historical considerations. In: Charney JI, Alexander LJ (eds) International maritime boundaries. Martinus Nijhoff, Boston

Oxman B (1996) Caspian Sea or lake: what difference does it make?. Caspian Crossroads Mag 1 (4):14

Wolfrum R (1990) Die Umsetzung des Seerechtsübereinkommens in Nationales Recht. United Nations (б.м.)

Wooldridge F (1992) Contiguous zone. In: Bernhardt R (ed) Encyclopaedia of public international law. North-Holland (б.м)

Zonn I, Zhiltsov S (2004) Kaspijskij Region (The Caspian region). Springer, Moscow

Chapter 6
The Regime for the Use of Nonliving Resources in the Caspian Sea

6.1 Reserves of Nonliving Resources in the Caspian Sea

Considering the overall resource potential of the Caspian region, it remains undisputed that the secured oil reserves of the Caspian region are considerably smaller than the energy potential of the Middle East. The oil production of the Caspian littoral states accounts for 20.6% and gas production for 26.2% of world reserves (BP 2019). The Caspian oil and gas industry is developing most especially in Azerbaijan (the Balahani–Sabunchi–Ramani site, the offshore Shah Deniz field, and the Azeri–Chirag–Gunashli field), Kazakhstan (Tengiz, Karachaganak, and Kashagan), and Turkmenistan (the South Yoloten–Osman field).[1]

Azerbaijan exports its natural gas mainly via the South Caucasus Pipeline (SCP), which is destined for Turkey, and since 2020 even directly to Europe.[2] Iran's reserves in the Caspian Sea are considerable. The main Caspian Sea port of Iran is Neka,[3] which facilitates swap agreements with Caspian Sea states.[4] The oil fields of Kazakhstan—Kashagan, Tengiz, and Karachaganak—are vast. The two latest are also natural gas fields, where much of gas is reinjected to increase oil production.[5] Turkmenistan is the leader in natural gas production in the Caspian Sea. It exports

---

[1] Aydin and Azhgaliyeva (2019) See (Farah 2015), pp. 179–193.

[2] Background Reference: Azerbaijan. U.S. Energy Information Administration. Updated Jan.7, 2019 https://www.eia.gov/international/analysis/country/AZE accessed 20 June, 2020.

[3] Iran Oil Swap (Neka, Tehran); - first oil transported in 2017, no update after that https://financialtribune.com/articles/energy/70265/neka-port-resumes-oil-swap-after-seven-years accessed June 20, 2020.

[4] Background Reference: Iran. U.S. Energy Information Administration. Updated Jan.7, 2019 https://www.eia.gov/international/content/analysis/countries_long/Iran/background.htm accessed 20 June, 2020.

[5] Background Reference: Kazakhstan. U.S. Energy Information Administration. Updated 20 June, 2019 https://www.eia.gov/international/content/analysis/countries_long/Kazakhstan/background.htm accessed June 20, 2020.

gas mainly to China via the Central Asia–China Pipeline (CACP) and also to Russia and Iran. Another route is the Turkmenistan–Afghanistan–Pakistan–India (TAPI) Pipeline, operating already since 2019.[6] Also, Russia is active in the oil extraction, mainly in the Northern Part of the Caspian Sea in the main fields in Kashagan and Filanovsky.[7]

6.2 International Legal Regulation of Nonliving Resources

The legal regime of the nonliving resources, according to the law of the sea, differs in respective categories of marine waters: firstly, in internal waters and territorial seas, where the coastal state exercises full and unrestricted sovereignty over nonliving resources above the seabed and in the subsoil of the submarine areas; secondly, in the economic zone and the continental shelf, where the coastal state exercises "sovereign rights" with respect to nonliving resources; and, thirdly, in the so-called Area, the seabed and ocean floor and subsoil thereof, beyond the limits of national jurisdiction, where nonliving resources are excluded from the sovereignty of individual coastal states and considered the shared heritage of humankind. This law of the sea division of maritime waters became the basis of the intergovernmental negotiations on the regime of nonliving resources in the Caspian Sea and has been adjusted appropriately in the Caspian Sea Convention in 2018. The consequences of the application of the law of the sea zones for the Caspian Sea will be analyzed in the next part, where prospects for a future legal regime of the nonliving resources are elaborated.

There are some special international legal models related to the use of transboundary nonliving resources, which are commonly used in the resource-related agreements. First is the "resource deposit clause," which is used rather rarely because of high legal risks. It requires states' cooperation in the case where at least one resource crosses the existing state boarder lines and at least one share of the resource field could be exploitable.[8] The second model of cooperation on shared resources is the so-called resource unitization. This model requires cooperation between parties, carrying, license or concession when their activities related to resources' use happen in the areas which overlap with each other. Other existing legal models for jurisdiction over the transboundary nonliving resources are as follows: "common management zones," "revenue sharing," "management cooperation," and "mutual restraint."

---

[6]Background Reference: Turkmenistan. U.S. Energy Information Administration. Updated July, 2016 https://www.eia.gov/international/analysis/country/TKM accessed July 2, 2020.

[7]Country Analysis Brief: Russia U.S. Energy Information Administration. Updated Oct.31, 2017 https://www.eia.gov/international/content/analysis/countries_long/Russia/russia.pdf accessed 20 June, 2020.

[8]UK–Norway, (Charney and Alexander 2003), pp. 1879 et seq. No. 9-15.

Last but not least is the model agreement of the "joint development" system, which is another intergovernmental model agreement, one that is usually applied to oil and gas fields and could be suitable for the regulation of the Caspian Sea. It regulates the exploration and exploitation of nonliving resources that have a transboundary character or are located in an overlapping zone.[9] There is only one famous case where the "joint development model" was used for a resource field that was located next to the coast of one of the contracting states without interfering into the adjacent area of the other contracting state.[10] The goal of the "joint development model" is to identify differing state interests that might arise from the demarcation and allocation of resources. Therefore, this model might be applied even before the final settling of a border agreement to enable early exploration and exploitation of the natural resources.[11] It is possible then that a subsequent delimitation can be performed respectively, one that takes in mind the situation achieved after the "joint development system" has been applied.[12] This may result in a final demarcation in zigzag form to avoid touching upon existing fields and saving the integrity of the already granted concessions.[13]

6.3 Claims on the Rights to Use Nonliving Resources in the Caspian Sea

The industrialization of the Caspian Sea is particularly evident in the growing number of artificial islands and other installations for the extraction of nonliving resources therein. The first of many facilities designed for resource extraction was the artificial island built east of the Kashagan oil field by Agrippa KKO Company.

The coastal states' conflicting territorial claims, which result in an overlapping of zones, speaks for the delimitation of the Caspian Sea. It shall, however, be considered that the requested territorial rights might injure the rights—also on the use of resources—assented to by states not involved in the delimitation process of the maritime zones in question.[14] This becomes particularly acute in the case of maritime areas, where the distance between the opposite shores is less than 400 nm, as is the case for the Caspian Sea. Since the end of the 1990s, a conflict has ensued between Azerbaijan and Iran regarding the Araz–Alov–Sarq fields in the south of the Caspian Sea. In July 2001, military threats were made by Iranian ships against British Petroleum, which was conducting seismic surveys at the Araz–Alov–Sarq

---

[9]Japan–South Korea 1974, ibid, pp. 1057 et seq. No. 5-12.
[10]Bahrain–Saudi Arabia, ibid, pp. 1489 et seq. No. 7-3.
[11]Canada–USA (Gulf of Maine): ibid, pp. 401 et seq. No. 1-3.
[12]Bahrain–Saudi Arabia: ibid, pp. 1489 et seq. No. 7-3.
[13]Trinidad–Tobago–Venezuela: ibid, pp. 675 et seq. No. 2-13(3).
[14]Aegean Sea, ICJ Rep. 1978, § 85, p. 35; Tunisia vs. Libya, ICJ Rep. 1982, § 75, pp. 61 and 62; Gulf of Maine, ICJ Rep. 1984, §195, p. 327.

fields on behalf of Azerbaijan. Another disputed area is the oil field Serdar/Kyapaz located within the overlapping zone between Azerbaijan and Turkmenistan. Turkmenistan also denies the right of Azerbaijan to exploit the oil fields of Chirag, Azeri, and Kyapaz/Serdar. Although both states are in agreement that the delimitation of the Caspian Sea bed should be conducted according to the principle of the median line, there is no agreement on where exactly the line shall run. The clashing legal claims of several states over the same maritime areas was regulated—even if only partially—in the northern part of the Caspian Sea through a set of Agreements between Azerbaijan, Kazakhstan, and Russia. Clashing legal claims are also present in other parts of the Caspian Sea.

Agreements reaching back to the nineteenth century reflected a lack of regulation of maritime borders in the Caspian Sea, which was continued in the Soviet–Iranian legal practice from 1921 to 1940, and thus contributed to the lack of clarity regarding the currently existing rights of the riparian states on the use of the natural resources. From the Soviet–Iranian theory of closed sea, which used to be applied to the Caspian Sea by the former littoral states, one can conclude that all Caspian resources were fully under the sovereignty of both riparian states. Another significant feature of the Soviet–Iranian legal doctrine of closed sea was the assumption that in the case of absence of another agreement, the littoral states shall exercise their sovereignty within the territorial waters and the regime of the central parts of the basin shall be the same as the regime of the high sea.[15] The treatment of the Caspian Sea as a closed sea was expressly confirmed in 1955 in Article 2 of Iran's National Law on Exploration and Exploitation of the Continental Shelf of 1949.[16] According to this article, one can conclude that the use of Caspian nonliving resources within the territorial waters of the Soviet Union and Iran was limited to the respective coastal states, but in the central part of the Caspian Sea, the exploitation of the resources was open to both. In 1970, the Oil Industry Ministry of the USSR divided the use of the Soviet part of the Caspian Sea among four Soviet republics—Azerbaijan, Kazakhstan, Russia, and Turkmenistan.[17] The internal zones of use created in the Soviet part of the Caspian Sea were determined according to the principle of the center line.

The lack of clarity with regard to the scope of the individual rights of the riparian states on the nonliving resources of the Caspian Sea grew in importance after the dissolution of the Soviet Union, which triggered unilateral attempts at regulation. In a communique of April 1996 between Kazakhstan and Russia, the latter recognized the rights of all Caspian riparian states to conduct all possible activities in the field of mineral and biological resources. Also, in a joint statement from October 1996, Kazakhstan and Azerbaijan guaranteed each other the right to exploit the natural resources of the Caspian Sea.[18] Both states recognized the need to define the states'

---

[15]See: (Buttler 1971), pp. 116–133.
[16]National legislation and treaties relating to the law of the sea (1974), XXXIV, p. 151.
[17]See Sect. 5.3.
[18]UN Doc. A/51/529 from October 21, 1996 (Azerbaijan–Kazakhstan).

sovereign rights in respective sectors in the Caspian Sea, and this was expressed also in the bilateral statement of Kazakhstan and Turkmenistan in February 2007.[19]

During the November 1996 meeting of the foreign ministers of five Caspian states in Ashkhabad, three of them (Russia, Iran, and Turkmenistan) signed a Memorandum on the joint use of the Caspian's natural resources.[20] According to this Memorandum, a tripartite company for the investigation and development of hydrocarbon resources was to be established—however, this did not come to pass. Azerbaijan and Kazakhstan refused to join this declaration. Russia's original concept of introducing a joint development area over the Caspian seabed beyond a 45-mile economic zone was rejected by the other riparian states. Therefore, Russia's recognition of unilateral actions concerning the Caspian Sea was finally expressed only with the signing of the North Caspian Agreements, providing for gradual progress toward a consensus-based solution to the Caspian status problem with the settlement, first and foremost, of the issues of exploitation of mineral resources and the environment, as well as fishing and navigation.[21]

In the period from 1998 to 2004, Russia, Kazakhstan, and Azerbaijan signed the so-called North Caspian Agreements, where in addition to sector delimitation, the regime of exploitation of the natural resources in the northern part of the Caspian Sea between these countries was settled. The treaties included clear provisions on the use of the nonliving resources in the northern Caspian Sea, although neither Iran nor Turkmenistan is a party to them. The Agreement of 1998 between Kazakhstan and Russia (Article 2) provides for the states' sovereign rights over the seabed and subsoil of the northern part of the Caspian Sea, particularly with regard to exploring and exploiting resources within the sectors' limits. States agreed on exclusively common rights to explore and exploit resources that extend across the median line set by the treaty. Detailed regulations on the potential joint exploitation works provide an additional protocol to the treaty. It covers two geological structures: Kurmangazy (Kulalinskaya) and Central Well, as well as a field called Kwalunskoye. Kazakhstan exercises sovereign rights over the Kurmangazy structure (Article 2), but Russia is entitled to take part in the exploitation in the form of a joint development (Article 3). Respectively, Russia exercises sovereign rights over the Central Well structure and Kwalunskoye field (Articles 4 and 5); however, Kazakhstan remains entitled to participate in the resources' exploitation (Article 4). The protocol provides also that in the case of the discovery of new transboundary geological structures in maritime boundaries, the states parties shall re-sign a new contract that would determine the economic activities in such an area.

In 2001, a similar agreement on the bottom of the northern part of the Caspian Sea was signed between Azerbaijan and Kazakhstan. It provided (Article 3) for the

---

[19]UN Doc. A/52/93 from March 17, 1997.
[20]See: (Mamedov 2001), p. 237.
[21]UN Doc. A/58/719–S/2004/137 vol. February 23, 2004 (Letter dated February 19, 2004 from the Permanent Representative of Azerbaijan and the Russian Federation to the United Nations addressed to the Secretary General).

parties' sovereign rights over the seabed and subsoil of the Caspian Sea, particularly with regard to exploring and exploiting the resources within the limits of the two settled national sectors. According to the 2003 Additional Protocol to this treaty, in the case of the identification of new fields extending across the median line between both national sectors, the states parties shall conclude a separate agreement.

The third of the Northern Caspian Agreements was concluded between Azerbaijan and Russia in 2002. It provides (Article 2, Para. 1) that the states parties shall exercise their sovereign rights over nonliving resources and other legitimate economic activities related to the exploration and exploitation of the resources within the treaty's defined sectors of the Caspian seabed and subsoil. The exploration and exploitation of the resources extending beyond the median line between both national sectors are to be performed by an organization authorized by states parties' governments according to the respective international legal practice (Article 2, Para. 2).

On December 2, 2014, the Agreement between the Republic of Kazakhstan and Turkmenistan on the delimitation of the bottom of the Caspian Sea between the Republic of Kazakhstan and Turkmenistan was signed and entered into force on July 31, 2015. This Agreement provides for the sovereign rights of Kazakhstan and Turkmenistan over the established sectors of the seabed and subsoil of the Caspian Sea for the exploration, development, and exploitation of resources; laying of pipelines and cables; building of artificial islands, berms, dams, overpasses, and platforms, as well as other engineering construction; and conduct of other legitimate economic activities.

The North Caspian agreements also guaranteed that their conclusion does not prevent a comprehensive agreement over the legal status of the Caspian Sea by all five littoral states. Accordingly, these treaties shall not predetermine the final demarcation of borders in the northern part of the Caspian Sea, but they shall reflect the states' claims on the delimitation of the Caspian Sea for the use of Caspian nonliving resources.

6.4 Use of Nonliving Resources in the Caspian Sea Reflected in the Multilateral Negotiations Among Caspian States

The littoral states' varied legal positions with regard to the legal status of the Caspian Sea envisaged in the multilateral negotiations from late 1990s and reflected in the draft of the Caspian Status Convention served as evidence of the existence of immense differences in their views with respect to the scope of each state's rights on the use of the nonliving resources of the Caspian Sea. Particularly controversial was the question on the recognition of the area of the territorial sea, including its seaward extension; the division of its seabed and subsoil; and the method of delimitation. The reason for this is that in the area of the territorial sea, similarly as in the state's internal waters, a coastal state possesses sovereignty, entitling it to extensive rights over the resources of this area. As every zone of the territorial sea may be subject to the sovereignty of only one coastal state, their delimitation contributed to great controversies and tensions between the riparian states. The provisions of the draft Caspian status agreement regarding the spatial order of the Caspian Sea revealed two irreconcilable concepts concerning its future legal status and regarding the existence of sovereign rights with respect to the resources of the Caspian basin. Recognition of the sovereign rights of states in the Caspian Sea is based on the concept of territorial sea. Through this, they would recognize the complete sovereignty of all Caspian littoral states in respective coastal seas, as well as national sovereignty over the resources. This position recognizing states sovereignty in the Caspian Sea was however rejected by some coastal states, which has been advocating for introduction of maritime zones with the status similar to the so-called "contiguous zone" known from the law of the sea. This implied a rejection of the coastal states' sovereignty over any area of the Caspian Sea, as well as of the rights to the natural resources.

According to the draft agreement on the Caspian status, which was worked out within the interstate negotiations, all Caspian states, except for Russia, agreed that coastal states shall be able to exercise their sovereignty over respective Caspian Sea maritime territories. Among the four Caspian coastal states, which were in favor of recognition of the state sovereignty in the Caspian Sea, there were different in defining this scope of sovereignty within particular maritime zones. According to Azerbaijan, Kazakhstan, and Turkmenistan, *sovereignty should extend to the air space over the Territorial Sea as well as the seabed and subsoil of the Territorial Sea* (Article 6(7), Para. 2). Iran claims that *sovereignty [should] be extended only to the waters superjacent to the seabed of the sector.*

Iran, in its position, proposed introduction of so-called "national sectors", where the rights of coastal states would be limited to development and use of maritime resources as well as implementation of other economic activities at the seabed and subsoil of the Caspian Sea. Such position rather echoes Iran's rejection of the sovereignty concept that was to be applicable over the national sectors in the Caspian Sea.

Regarding installations and structures in the Caspian Sea, the states agreed in the Draft Caspian Status Agreement that *State Parties shall exercise their sovereignty [Russia proposes to remove this term and replace it with the term "jurisdiction"] over their nationals, ships in their ownership and over installations and structures in the Caspian Sea according to the norms of international law* (Article 12 (11)). Such a norm shall be exercised similarly as foreseen in the concept of territorial Sea and exclusive economic zone in the law of the Sea. In the territorial Sea a coastal state has unlimited sovereignty including over installations and structures, when in the exclusive economic zone coastal states exercise sovereign rights over the extraction of natural resources.

Such regulation of the legal regime of the seabed and the subsoil of the Caspian Sea seaward of the Territorial Sea and the Fishing Zone or the Zone of National Jurisdiction would not correspond with these legal standards recognized under the law of the sea. The Draft Caspian Status Agreement in Article 5(6) 2 provided that "The seabed and its subsoil are to be separated for the purpose of exercising the rights for the extraction of mineral resources as well as other legitimate socio-economic activities concerning the development of the resources of the soil and subsoil" (Iran refuses). No special legal regime for the Caspian areas seaward of the areas covered by the exclusive coastal states' rights have been understood as equipping the seabed and subsoil maritime sectors of the Caspian Sea with equal status. This differs from the UNCLOS regulations, which recognized the "Area" located beyond the limits of national jurisdiction as a shared heritage of humankind and where no sovereign rights may be acquired or exercised.[22]

6.5 New Regulations for the Use of Nonliving Resources in the Caspian Sea According to the Caspian Sea Convention 2018

The Caspian Sea Convention adopted in 2018 reflects the legal regime of the use of the nonliving resources in the Caspian Sea. In Article 2, Para. 2, the Convention confirms the rights and obligations of the parties in respect of the use of the Caspian Sea, including its natural resources as well as seabed and subsoil. The regulation of the nonliving resources differs between the maritime zones, which either remain under the sovereignty of the Caspian Sea countries or beyond it. In the first case, the sovereignty of each party to the Convention has been recognized in the so-called internal waters and adjacent so-called territorial waters, as well as the seabed and subsoil thereof (Article 6). According to the Caspian Sea Convention (Article 7, Para. 3), the delimitation of the internal and territorial waters between the states with adjacent coasts shall be affected by an agreement between those states. It is still a work in progress and will certainly take the coastal states some time to conclude such agreements. If zones of internal and territorial waters would be legally recognized, the coastal states would poses their exclusive sovereignty over nonliving resources located in these zones.

In respect of the regime of the nonliving resources of the seabed and subsoil beyond the territorial waters, the Caspian Sea Convention is, however, not forthright. These parts of the Caspian Sea seabed and subsoil shall be delimited into sectors. Within the frame of these sectors, states are able to exercise their sovereign rights to subsoil exploitation and other legitimate economic activities for the development of the resources of the seabed and subsoil (Article 8, Para. 1). Further, the Caspian Sea Convention states:

> The exercise of sovereign rights of a coastal State under Para. 1 of this Article must not infringe upon the rights and freedoms of other Parties stipulated in this Convention or result in an undue interference with the enjoyment thereof (Article 8. Para. 4).

---

[22]Art. 1, §. 1 UNCLOS, Arts. 135 and following. Many of UNCLOS's provisions regarding the "Area" were amended by Agreement relating to the Implementation of Part XI of the United Nations Convention on the Law of the Sea of 10 December 1982.

The states' activities, among others related to the exploration and exploitation of resources within the sectors of the Caspian Sea's seabed and subsoil, shall be conducted in accordance with other agreements between its littoral states, consistent with the Convention (Article 4). The Convention recognizes therefore a binding force of already existing agreements concluded previously between the parties for the purpose of exploration and exploitation of the seabed and subsoil resources. Such agreements could offer a lawful basis for the conduct of activities in the Caspian Sea for the purpose of exploration and exploitation of the seabed and subsoil within relevant sectors. Also, Article 20 of the Caspian Sea Convention says:

> This Convention shall not affect rights and obligations of the Parties arising from other international treaties to which they are parties.

On one side, it is to be stressed that not all littoral states have adopted seabed delimitation agreements prior to the signing of the Caspian Sea Convention. A possible negotiation on the future demarcation of some sectors justifies concluding tacit agreements on a preliminary settling of sector boundaries. On the other side, agreements were concluded in the past to divide the North Caspian Sea into proportional national sectors and guarantee to the states parties the sovereign right to the use of the natural resources within these sectoral borders. They were concluded by Azerbaijan, Kazakhstan, and Russia, and later Turkmenistan, and entered into force prior to the designing of the Caspian Sea Convention. A reading of the Caspian Sea Convention's Article 8, Para. 1, and Article 8, Para. 4, together with Article 4, allows for the assumption that the states parties of the Caspian Sea Convention have recognized the continuing binding force of the sectoral delimitation conducted according to the North Caspian Agreements.[23] Such interpretation can be reinforced by the fact that after the conclusion of the Caspian Sea Convention in 2018, there were no further official governmental statements from the states that did not participate in the North Caspian Agreements that would raise relevant concerns about their legally binding force.

6.6 Conclusion

Rights on the use of nonliving natural resources in maritime space offer an important added value to states' economic development and may oft be seen as a reason for countries' claim on the conduct of delimitation of maritime areas. The international law of the sea defines the scope of states' rights to the exploration and exploitation of nonliving natural resources in maritime zones. Defining the states' rights is particularly challenging in case of resources covered by overlapping claims by the coastal states, but also some legal methods of delimitation are available in the international

---

[23] Dissenting opinion: Delimitation agreements might need to be amended in case when they affect interests of third states, Kadir (2019), p. 399.

praxis. Until the conclusion of the Caspian Sea Convention in 2018, the legal regime of the abounding nonliving resources in the Caspian Sea, especially oil and gas, was very difficult to resolve.

In the time of the Soviet–Iranian control over the Caspian Sea, the use of natural resources was conducted based on the concept of *mare clausum*. In this state praxis, the Caspian space was informally divided into two separate zones exclusively used by each state. After the dissolution of the Soviet Union when debating about possible delimitation of the states' rights on the Caspian nonliving resources, the parties were referring to the legal concepts of the Caspian Sea as a sea, lake or condominium. And later on, from the moment of the conclusion of the North Caspian treaties, there were two directions in the development of the interstate negotiations: first, the conclusion of bi- and trilateral agreements on the use of the northern parts of the Caspian seabed and, second, multilateral negotiations undertaken in the form of a future Convention on the legal status of the Caspian Sea. The first was contested by the remaining costal states; however, they undertook other unilateral measures themselves. The latter shall have defined future maritime zones in the Caspian Sea and respectively the scope of states' rights within these zones. The conclusion of the Caspian Sea Convention in 2018 offered a framework regulation for the legal regime of the use of natural resources. The Caspian Sea soil and seabed and its natural resources were subjected to two different legal regimes. First, the so-called internal waters and territorial waters and the natural resources within these zones have been fully subjected to the national sovereignty of the relevant coastal state. The coastal state can exercise all rights related to the development, use, and protection of the natural resources. Beyond these zones, the Caspian Sea seabed and subsoil are to be in future divided into sectors according to other separate agreements between the states with adjacent and opposite coasts. The exercise of sovereign rights to subsoil exploitation and other limited economic activities related to the development of the resources of the seabed and subsoil must not infringe upon rights and freedoms of other states parties. In this context, the Caspian Sea Convention is quite clear about the recognition of the legal regime for the delimitation of the seabed and subsoil and the regime of the natural resources to be included into the so-called North Caspian Agreements between Azerbaijan, Kazakhstan, Russia, and later Turkmenistan. After the conclusion of the Caspian Sea Convention, there have been no more official objections from Iran as to the recognition of the provisions of the North Caspian Agreements for the regime of delimitation and use of natural resources within sectors shared by states parties to the North Caspian Agreements. Still, the borders of maritime sectors between Iran and Azerbaijan and Turkmenistan shall be defined in separate agreements in the future to come.

References

Aydin U, Azhgaliyeva D (2019) Assessing Energy Security in the Caspian Region: The Geopolitical Implications for European Strategy. ADBI Working Paper (s.l.)

References

BP (2019) Statistical Review of World Energy 68th edition. pp 17 et seq. [Online] Available at: https://www.bp.com/content/dam/bp/business-sites/en/global/corporate/pdfs/energy-economics/statistical-review/bp-stats-review-2019-full-report.pdf. Accessed 18 July 2020

Buttler W (1971) The Soviet Union and the law of the sea. John Hopkins Press, Baltimore

Charney J, Alexander L (eds) (2003) International maritime boundaries. Martinus Nijhoff, Dordrecht

Farah P (2015) Energy security, water resources and economic development in Central Asia. In: Rossi P, Farah PD (eds) Energy: policy, legal and social-economic issues under the dimensions of sustainability and security, World Scientific Reference on Globalisation in Eurasia and the Pacific Rim. Imperial College Press & World Scientific Publishing, London, pp 179–193

Kadir RA (2019) Convention on the legal status of the Caspian Sea. ILM 58:399

Mamedov R (2001) International legal satus of the Caspian Sea; issues of theory and practice. Turkish Yearb Int Relat 32:217–259

Chapter 7
The Legal Regime of the Living Resources of the Caspian Sea

7.1 Tensions Between the Protection of Fish Stocks and the Oil Industry in the Caspian Sea

Extensive commercial fishing and the rapidly growing number of oil-related offshore activities, along with the rapid development of new technologies, contribute to the increasingly unwanted intrusion into the traditional areas of the Caspian Sea, such as fishing. There is a biological limit to the exploitation of fish reserves, which is determined by the "Maximum Sustainable Yield," i.e., the largest catch to be taken from the stocks without destroying them.[1] The establishment of protection zones, the construction of substantial anchoring systems, as well as the abundance of mining equipment in the areas with the best fish stocks cause serious problems for fishing. Therefore, an important legal objective for legal regimes regulating use of the Caspian Sea, both on the transboundary as well as national levels of regulation, is to develop a new set of regulations able to reconcile the need of traditionally versatile usages of the Caspian Sea's—nonliving resources, its fishing and shipping regimes.

A fundamental concern is the Caspian fish stocks, which are characterized by a large quantity and variety. The Caspian Sea is inhabited by 153 species of fish, of which about 30 species are of high economic value.[2] According to the Caspian Environmental Programme, up to 500,000–600,000 tons of fish is caught in the Caspian Sea annually, with the majority being beluga (*Huso huso*), sterlet (*Acipenser ruthenus*), and migratory marine species of herring, pike perch (*Stizostedion*), common carp (*Cyprinus carpio*), common bream (*Abramis brama*), catfish (*Silurus*

---

[1] See: (Anderson 1975), pp. 159 et seq.
[2] Species in Caspian Sea. Fishbase. See: https://www.fishbase.de/TrophicEco/FishEcoList.php?ve_code=154 last modified 21/01/2020 (accessed 24 June 2020); See: (Zonn and Zhiltsov 2004), p. 43.

glanis), and the Caspian roach (*Rutilus caspicus*). Of these species, the ones with the largest commercial value are sturgeons (five species) and the roach (three species).[3]

Until 1991, two Caspian coastal states—the Soviet Union and Iran—had the control of the entire caviar market and were responsible for the conservation of fish stocks. With the disintegration of the USSR, however, state protection was limited and the exploitation of resources was furthered by poaching. In 1997, the Conference of the Parties of the Washington Convention on International Trade in Endangered Species of Wild Fauna and Flora (CITES), which all Caspian littoral states except Turkmenistan are parties to, included all sturgeon species in Appendix II, which lists the species threatened with extinction. Thus, since 1997, the exporter states of these endangered species have had the duty to abide by the strict regulations of CITES, including the necessary approval and special designation systems (Article VIII CITES). This implicated the Caspian states' obligation to enforce the CITES provisions regarding the introduction of systems of catch approvals as well as special designation systems to protect species. States are obliged to prohibit the trade of endangered species.[4]

According to the Secretariat of CITES, the Caspian Sea provides approximately 90% of world caviar stocks. The official annual catch has fallen from 30,000 tons in the late 1970s to less than 10,000 tons at the end of the 1990s.[5] As caviar is a popular local delicacy, Caspian Sea countries must also make efforts to control domestic trade in sturgeons. So far, the states have not been successful in stopping poaching and are far from responsible fisheries management. The required bans on fishing in areas with overfishing, however, are difficult to enforce over extended periods. The poaching of valuable fish stocks, particularly of sturgeons, is very common. Application of selective fishing methods which allowed avoiding the undesirable catch as well as avoiding catch of immature fish are expensive.[6] After years of interstate negotiations in the Caspian region, the problems of living resources have been resolved finally in 2018 with the adoption of the Caspian Sea Convention.

The negotiations on preliminary national allowances for caviar production used to always be a point of contention in the framework of the International Commission on Aquatic Resources of the Caspian Sea (ICARCS). The Commission of the Caspian Sea is the CITES regional body responsible for the allocation of the total allowable catch (TAC) among Caspian states. The authorization for national quotas by the CITES is based on the Conservation Action Plan for the Caspian Sea Sturgeon Fisheries concluded by the riparian states during the 45th CITES Standing

---

[3]Caspian Environment Programme. 3 Sept 2013 See: https://archive.is/20010712173449/http://www.caspianenvironment.org/biodiversity2.htm (accessed 24 June 2020).

[4]Sturgeons and the causes of their extinction. See: https://wwf.ru/en/what-we-do/seas/sturgeon-and-the-reasons-for-their-disappearance/#:~:text=All%20countries%20of%20the%20Caspian,too%20high%2C%20is%20actually%20priceless.

[5]'Caspian Sea states to resume caviar trade' 6 March 2002 http://www.cites.org/eng/news/pr/2002/020306_caviar_resumption.shtml accessed July 2, 2020.

[6]Caspian Basin: No Way to Halt Sturgeon Poaching May 8, 2008 https://eurasianet.org/caspian-basin-no-way-to-halt-sturgeon-poaching accessed July 2, 2020.

Committee Meeting in 2001. From 2001 the CITES's regulations for the Caspian domestic trade and markets oblige states to issue 12-month action plans and to provide for their commitment, to submit detailed caviar production data. As caviar stocks continued to decline throughout the 1990s, the parties to CITES decided to include all sturgeon species—including the rapidly declining beluga—in CITES' Appendix II, which restricts their trade on a scientific basis but does not ban it entirely; trade is illegal only for species listed under Appendix I of CITES. Respectively, all exports of caviar and other sturgeon products have had to comply with strict CITES provisions. The amount of sturgeon that can be harvested (TAC) by the Caspian coastal states is based on two not entirely compatible stock and catch assessment methods: sample trawling used by the former Soviet republics and catch-per-unit-effort (CPUE) used by Iran.

Caspian states, despite using different TAC methods, still cannot counteract illegal fishing. Because of the problem on caviar, CITES introduced a temporary ban on the export of caviar and other sturgeon products in 2001. It forced the coastal states to improve the regional situation in the region. Still, the CITES Secretariat was not able to publish the annual export quotas for 2002–2005. Also in later years (2006, 2009, 2012) the Caspian coastal states did not provide sufficient information about the sustainability of their sturgeon catch and were not given quotas for commercial sturgeon fishing. In August 2013, Russia, supported by some other riparian states, suggested to introduce a five-year moratorium on sturgeon catching in the Caspian Sea, which could help restore the sturgeon population to a commercially sufficient level. The moratorium was again extended until the end 2020. CITES, having defined species to be protected, sets up the export quotas in an annual resolution of the Conference of Parties for one calendar year. According to the annual Notification to the Parties by the Secretariat in 2020, national export quotas for six CITES-protected species from the sturgeon family found in the Caspian Sea accounts for zero.

7.2 Regime of the Living Resources in International Law

Living resources are not a coherent biological species category but can be divided into numerous marine products, such as conventional and unconventional products (e.g., deep-sea fish), or sea creatures used for medicinal purposes. Hence, the law of the sea regulation of the living maritime resources contains many different norms.[7] The basic rules of the legal regime of living resources is provided by United Nations Convention on the Law of the Sea (UNCLOS)[8] and certain instruments adopted after its entry into force, i.e., the Agreement on Implementation of the Provisions of the United Nations Convention of 10 December 1982 on the Conservation and

---
[7]See: (Kindt 1984), p. 9.
[8]See: (Hyvarinen et al. 1998), pp. 323–338.

Management of Straddling Fish Stocks and Highly Migratory Fish Stocks, the Food and Agriculture Organization's Code of Conduct for Responsible Fisheries, the complementary FAO Agreement to Promote Compliance with International Conservation and Management Measures by Fishing Vessels on the High Seas,[9] and the Global Programme of Action for the Protection of the Marine Environment from Land-Based Activities.[10] Also, the United Nations Framework Convention on Climate Change (UNFCCC) and the Paris Agreement deal indirectly with fishery resources, which face high risk because of increasing global temperatures.

According to the current law of the sea, the territorial sovereignty of a coastal state over its territorial sea enables the coastal state to reserve to its nationals all fishery rights in the area covering this space. On the other hand, the high seas are open to all states, whether coastal or landlocked. Freedom of the high seas includes, *inter alia*, freedom of fishing.[11] The right to engage in fishing on the high seas is, however, restricted by treaty obligations, the rights and duties occurring from UNCLOS, as well as the interests of coastal states. Such main duty is to take, or to cooperate with other states in taking, such measures for their respective nationals as may be necessary for the conservation of the living resources of the high seas (Article 117 UNCLOS). Between the areas covered by the states' full sovereignty over the living resources in their territorial sea and the area of freedom of fishing on the high seas, there was a need to create an additional zone recognizing states' fishing rights but simultaneously guaranteeing stock protection. Its initial form was the fishery zone, which was gradually replaced in state practice by the concept of exclusive economic zone.

The principle of freedom of fishing was originally established as a result of the principle of freedom on the high seas.[12] Nevertheless, states have tried to reserve certain exclusive rights outside of their territorial seas, particularly with regard to fishing and mining, which has also finally found recognition in international case law.[13] Already in the 1940s, some of the Latin American states raised claims regarding the establishment of special fishing zones (extended up to 200 nautical miles measured from the baselines). Their rationale was the need to guarantee food security for their population through natural resources.[14] As a result, merely sovereign rights on fishing were recognized.

---

[9]Code of Conduct for Responsible Fisheries (1995).

[10]UN Doc. A/51/116 (1996).

[11]Articles 87 and 116 UNCLOS.

[12]Articles 2 § 3 No. 2 of Convention on the High Seas of 1958, Article 1, Section 1 of the Convention on Fishing and conservation of the living resources of the high seas; Article 87, § 1 lit. e i.V.m., Article 116 of UNCLOS.

[13]ICJ Rep. 1973, pp. 44 f. over support Latin American claims regarding 200 miles economic zones, also ICJ Rep. 1974, p. 192 in the Fisheries Jurisdiction Cases (Germany/Great Britain). The recognition of states' rights on fish stocks located out of the Territorial Sea was so reflected in the ICJ decision in the case of so-called "Icelandic fisheries dispute."

[14]See: (Garcia-Amador 1974), pp. 33 et seq.

7.2 Regime of the Living Resources in International Law

The introduction of the concept of exclusive economic zones and respective restrictions on the freedom of fishing on the high seas led to the escalation of conflicts of fishing interests among the coastal states. This opened the high seas to the fishing states operating worldwide. Additional tightening of fishing rights on the high seas, following the growing dissatisfaction of fishery states, was introduced because of the steady downward trend of existing fish stocks. The states' far-reaching obligations to protect the living resources were established, based on, among others, the Convention on Fishing and Conservation of the Living Resources of the High Seas of 1958, Agreement Concerning Co-operation in Marine Fishing of 1962, Fisheries Convention of 1964, and Convention on Conduct of Fishing Operations in the North Atlantic of 1967.

The establishment of an exclusive economic zone (EEZ) was the greatest achievement of the Third Law of the Sea Conference (1973–1982), but it was first recognized as part of customary international law,[15] which was initially met with great difficulties.[16] Its adoption contributed to a substantial restriction of the principle of freedom of fishing on the high seas. Its maximal breadth shall not be more than 200 nautical miles measured from the baseline. The delimitation of the exclusive economic zone may be conducted between states with opposite or adjacent coasts based on an agreement guaranteeing an equitable and appropriate solution. Within the EEZ, the coastal states' sovereign rights to the exploration, exploitation, conservation, and management of the living and nonliving natural resources of the waters superjacent to the seabed and its subsoil are guaranteed.[17] The coastal states thus have no sovereignty over their own economic zone but merely sovereign rights with respect to the living and nonliving resources. The regulation of fishing and fish stocks is subordinated to the coastal state's sovereignty.[18] Every coastal state shall ensure the application of appropriate conservation and management measures in order not to jeopardize the existence of the living resources in the exclusive economic zone due to overexploitation and to preserve or bring back the populations of harvested species to levels securing the maximum sustainable yield.[19] In its authority regarding the exploitation of living resources, the coastal state shall promote the objective of optimum utilization of the living resources in the exclusive economic zone.[20] For this purpose, it shall determine its capacity to harvest the living resources in the exclusive economic zone. Where the coastal state does not have the capacity to harvest the entire allowable catch, it shall, through agreements or other arrangements, give other states access to the surplus of the allowable catch.[21] According to the law of the sea, the special rules extend over stocks

---

[15] Case Concerning the Continental Shelf, ICJ Rep. 1982, 74.
[16] See: (Hafner 1987), p. 185.
[17] Article 56, § 1 a) UNCLOS.
[18] Article 61, § 1 UNCLOS.
[19] Article 61, § 2, 3 UNCLOS.
[20] Article 62, § 1 UNCLOS.
[21] Article 62, § 2 UNCLOS.

occurring within the exclusive economic zones of two or more coastal states or both within the exclusive economic zone and in the areas beyond and adjacent to EEZ. It concerns for instance highly migratory species, marine mammals, anadromous stocks, catadromous species, and sedentary species.[22]

Historically, the exclusive economic zones used to be fishery zones, but gradually the latter was replaced by the former. However, still many coastal states worldwide establish fishing zones among their maritime areas instead of an exclusive economic zone, though UNCLOS does not include any direct reference to it. Originally, the fishing zone concept was understood as covering merely up to a maximum breadth of 12 nautical miles measured from the baseline.[23] It recognized preferential but nonexclusive fishing rights of the states that are dependent on coastal fisheries in a particular way.[24] State practice in the 1970s confirmed the success of the fishing zone regimes. The exclusive fishing zone is a part of the exclusive economic zone. With regard to the fishery zone, all relevant provisions of UNCLOS on exclusive economic zones can be applied by analogy. In the fishery zone, the rights and jurisdiction of the coastal state are limited exclusively to living resources. The fundamental difference between a fishery zone and an exclusive economic zone is that in the fishery zone, the exercise of individual coastal states' sovereign rights is limited merely to the exploring, exploiting, conserving, and managing of the living resources.[25]

Rights on the use of natural resources may be allocated following different approaches for instance of resources' utilization, profit-sharing, legal management, etc. Territorial sovereignty is the only concept in international law that regulates the equitable distribution of states' rights to the use of resources.[26] There is no exclusive, universally binding legal model for the use of natural resources. Some standards for assigning fishery usage rights can be found in individual interstate arrangements on fishing quota allocation as they define the allocation of rights between particular states and private companies.[27] A suitable example could be the rules on fishery management outside the 200-mile zones adopted by the regional fishery organization North-East Atlantic Fisheries Commission (NEAFC).

As a result of the introduction of the exclusive economic zone, a new category of states' rights with respect to transboundary fish resources has emerged. Whether it is possible to reconcile the diverging state interests by virtue of the adoption of cooperation agreements depends on numerous legal[28] and nonlegal (biological,

---

[22]Corresponding: Articles 63–68 UNCLOS.

[23]Fisheries Jurisdiction Case, ICJ Rep. 1974, p. 192. Critics: (Churchill 1975), p. 82.

[24]ICJ Rep., 1974, p. 196.

[25]Handbook on the Delimitation of Maritime Boundaries (2000), p. 9.

[26]Exceptions for restricted multilateral treaties example respect of the High Sea, Spitzbergen, Antarctica.

[27]See: (Hafner 1987), pp. 119 et seq.

[28]See: Case Law: Lac Lanoux Arbitration (1957). In: 24 International Law Reports 1957, p. 101; North Continental Shelf Cases (1969) in: ICJ Rep. 1969; Case Concerning the Delimitation in the Gulf of Maine Area (1984) in: ICJ Rep. 1984; Fisheries Jurisdiction Cases (1974) in: ICJ Rep. 1974. UNCLOS: Articles 2, 56, § (1)(a), 77, § 77(1), 116, 117, 63, § (1), 63(2), 65–67.

technological, economic, social, and political) factors.[29] The legal regime of fish stock management varies depending on the location of the fish sock. First type concerns stocks occurring within the exclusive economic zones of two or more coastal states or both within the exclusive economic zone and in the area beyond and adjacent to it. The second applies to fish stocks located within areas with overlapping claims of two or more coastal states (exclusive economic or fishery zones), which thereby hampers the delimitation of the zones.[30]

An additional problem regarding the fishery regime arises from the partially contentious relationship between respective provisions on the regimes of the exclusive economic zone and the continental shelf under customary international law. The question is whether the states' sovereign rights on fishing within their exclusive economic zones are impacted by the special regimes of the fishery safety zones adopted for certain oil or gas exploration fields, and this question should be answered positively. Such a conclusion seems to be legitimate in the light of current state practice. Respectively, fishing in such safety zones is subjected to the legal regime of the continental shelf and not to that of the exclusive economic zone.[31]

The always contentious issue of quota on export is not specifically regulated by law; neither does CITES require member states to establish quota to ensure limited trade in CITES-listed species. However, this highly effective practice for the regulation of international trade in wide flora and fauna has been, however, recognized in CITES COP Resolution 14.7 on Management of Nationally Established Export Quotas 2006. Each state is individually responsible for setting national export quotas for CITES-protected species.

7.3 Historical Development of Regulations of Fishing in the Caspian Sea

7.3.1 Soviet–Iranian Fishery Regulation

The original rights of Caspian coastal states to the exploitation of fishery resources were settled in the Friendship and Cooperation Treaty between the Russian Soviet Socialist Republic (RSSR) and Persia of 26 February 1921, wherein Persia explicitly recognized the great importance of the Caspian fisheries for Russia's food supply (Art XIV). In October 1927, the Agreement regarding the exploitation of the fisheries on the southern shore of the Caspian Sea, with additional protocol, was signed. It stipulated that commercial fishing outside coastal zones 10 nautical mile wide was restricted exclusively to a common Russian–Persian company 50% owned by each coastal state (Article 5). This Agreement was originally concluded for a

---

[29]Hey (1987), pp. 15 et seq.
[30]See: (Lagoni 1992).
[31]See: (Ulfstein 1998), pp. 237 et seq.

period of 25 years and not extended by Iran. The regulation of the Caspian states' fishing rights was extended in the Trade and Navigation Treaty concluded between the USSR and Iran on March 25, 1940. It determined a 10-nautical-mile coastal zone, within which each party enjoyed exclusive rights on fishing. Seaward of the fishing zone, both coastal states possessed unrestricted freedom of fishing. The principle of exclusivity and commonality of the coastal states' rights to living resources located outside the 10-nautical-mile exclusive fishing zones remained in force until the collapse of the Soviet Union. It was upheld also after the dissolution of the USSR, as reflected in the legal positions of the new coastal states.

7.3.2 Agreement on the Conservation and Rational Use of the Aquatic Biological Resources of the Caspian Sea

At the conference of October 4, 1992, the parties agreed upon a determination of the spheres of joint actions and organized six specialized committees; however, only the committee on biological resources carried out its works. The International Commission on Aquatic Resources of the Caspian Sea (ICARCS) was created by four littoral states and was not joined by Iran until 2003. Its goal was to regulate Caspian fisheries by defining the total allowable catch (TAC) and distributing between the coastal states the catch quota regarding major commercial fish species (sturgeons, kilka, seals). The states' quota system is based on the methodology of the Federal State Budget Scientific Institution Caspian Fisheries Research Institute and depends on their contribution to species reproduction (volume of freshwater inflow, number of fingerlings from natural spawning grounds, number of released fingerlings from hatcheries, habitat feeding grounds, and resources). The Commission actively works in the area of conservation and the use of Caspian bioresources, scientific cooperation, and data exchange to calculate distribution quotas between countries.[32]

The Commission, while meeting twice a year under a two-year rotating chairmanship from each country, prepared in 1993 a project for a convention on the use and protection of biological resources. At the meeting in Ashkhabad held between January 30 and February 2, 1995, the issue of the extension of the exclusive jurisdiction zones of the coastal states over fisheries remained unresolved. A consensus could not be achieved due to different state positions proposing 15 miles (Russia), 25 miles (Kazakhstan), 30 miles (Iran), and 40 miles (Turkmenistan and Azerbaijan).[33] The Agreement on the Conservation and Rational Use of Aquatic Biological Resources of the Caspian Sea was finally adopted in September 2014 and entered into force on May 24, 2016. The Agreement applies to the aquatic biological resources of the Caspian Sea (Article 1), which are defined as fish, shellfish, crustaceans, mammals and other species of aquatic animals and plants (Article 2).

---

[32]CAS State of Environment (2010).
[33]See: (Mamedov 2001), p. 237.

7.3 Historical Development of Regulations of Fishing in the Caspian Sea

The purpose of this Agreement is the conservation and rational use of the aquatic biological resources of the Caspian Sea, including the management of shared aquatic biological resources (Article 3). The principles of state cooperation for the implementation of this Agreement are the following: *(1) the priority of the conservation of aquatic biological resources of the Caspian Sea over their commercial use; (2) sustainable use of shared aquatic biological resources; (3) the application of generally accepted international rules acceptable to the Parties in relation to the regulation of fishing and the conservation of aquatic biological resources of the Caspian Sea; (4) conservation of the ecological system of the Caspian Sea and the biological diversity of aquatic biological resources; (5) the use of scientific research as the basis for the conservation of aquatic biological resources and the management of shared aquatic biological resources; (6) ensuring the compatibility of measures for the conservation, rational use of aquatic biological resources of the Caspian Sea and the management of shared aquatic biological resources throughout the range of species* (Article 4). The following are the areas of state cooperation: *(1) conducting coordinated research; (2) development of measures to regulate the harvesting of shared aquatic biological resources; (3) development of measures to combat illegal, unreported, unregulated fishing and illegal circulation of aquatic biological resources; (4) collection, provision and exchange of fishing statistics in a format agreed by the Parties; (5) the development and implementation of short, medium and long-term programs for the reproduction and conservation of joint aquatic biological resources and their habitats, including the release of juvenile sturgeon species; (6) development of recommendations on the use of fishing gear and fishing technologies for joint aquatic biological resources; (7) the exchange of scientific information and specialists, seminars, conferences and training courses* (Article 5).

The Agreement on the Conservation and Rational Use of Aquatic Biological Resources of the Caspian Sea foresees harvesting of joint aquatic biological resources. The national quotas are to be determined on the basis of common criterias established by the Quota Commission (Article 6, Para. 1). Joint aquatic biological resources include sturgeons, sprats, seals, as well as fish species included in the list by the Commission (Article 2). In case parties cannot agree on the quotas, the Quota Commission can make the final decision (Article 6, Para. 2). Coastal states shall conclude separate arrangements with each other in case they are not able to use their quotas within this scope of the Total Allowable Catch. Agreement defines also the geographical scope of its application as that *commercial fishing for sturgeon species is carried out in rivers and their estuaries, as well as in marine areas established by a decision of the Commission, taking into account the traditional fishing methods of each of the States of the Parties* (Article 7). Countries shall minimize by-catch of sturgeon species (Article 8). They *shall also take all possible measures to combat illegal unreported and unregulated fishes* (Article 9).

The Agreement established a special Quota Commission comprising representatives of each party. The Commission shall meet regularly, at least once a year, and its chairmanship shall be executed in turn by representatives of each state. Decisions of the Commission are adopted unanimously (Article 10). The Agreement defines the following powers of the Commission: *(1) coordinate the conservation, reproduction, and rational use of shared aquatic biological resources; (2) to amend the list of types of joint aquatic biological resources; (3) annually determine the Total Allowable Catches of shared aquatic biological resources and allocate them to national quotas; (4) establish criteria for the allocation of the Total Allowable Catch of shared aquatic biological resources to national quotas; (5) to coordinate the volumes of catch of the national quota of sturgeon species of fish in case of its transfer to another Party fishing in their rivers; (6) regulate the fishery and conservation of shared aquatic biological resources based on fishing restrictions, which may include the following measures: the prohibition of fishing in certain areas and in relation to certain types of aquatic biological resources for certain periods; seasonal closure of fishing in certain areas and for certain types of aquatic biological resources; establishment of the minimum size and weight of harvested (caught) aquatic biological resources; determination of the mesh size and design of the tools for extraction (catch) of aquatic biological resources; other agreed fishing restrictions; (7) consider the submitted reports on the development of quotas for the catch of shared aquatic biological resources and reports on the implementation by the Parties of other decisions of the Commission; (8) consider fishing statistics, including the number of vessels participating in the fishery, and their catch; (9) approve fisheries rules in relation to shared aquatic biological resources; (10) to develop and approve the necessary programs and projects for the protection of rare and endangered species of shared aquatic biological resources, the list of which is approved by the Commission, and their habitat; (11) to approve and coordinate the agreed research programs on joint aquatic biological resources, to establish the frequency of such work; (12) to collaborate with relevant specialized international organizations to achieve the objectives of this Agreement; (13) to contribute to the resolution of controversial issues in the field of conservation, reproduction and rational use of shared aquatic biological resources; (14) provide a procedure for the joint assessment of quantitative and qualitative characteristics of juveniles produced; (15) create working groups, determine the directions and programs of their activities; (16) monitor the implementation of decisions made; (17) establish export quotas for sturgeon species of fish and products from them; (18) carry out other functions and make other decisions that may be necessary for the implementation of this Agreement* (Article 11).

7.3.3 Regulation on the Living Resources of the Caspian Sea According to Other Regional Agreements

After the fall of the Soviet Union, riparian states supported the idea of a community approach to the use of Caspian living resources. This was reflected also in the bilateral agreements on the delimitation of the northern part of the Caspian Sea. In the Agreement between Azerbaijan and the Russia of 2002 (Article 1) and in the

Agreement between the Republic of Azerbaijan and the Republic of Kazakhstan 2003 (Article 1), the legal regime of the water column was deliberately left unregulated because of the claims of Azerbaijan in this regard. Also, the Agreement between Kazakhstan and Turkmenistan on the delimitation of the bottom of the Caspian Sea from 2015 did not refer to water column and respectively to its living resources. Meanwhile, the Agreement between Kazakhstan and Russia 1998 (Article 1) expressly provides that the water column remains in common use, even when the seabed and sea soil of the northern part of the Caspian Sea were delimited between the parties to this Agreement for the use of other resources.

The greatest achievement in the strengthening of the existing legal regime on living resources was the conclusion of the Tehran Convention.[34] The Protocol on Biodiversity to the Tehran Convention, which was adopted on May 30, 2014, remains in close thematic relations with the Agreement on the Conservation and Rational Use of the Aquatic Biological Resources of the Caspian Sea and in detail has been discussed in Chap. 10 of this book. The Protocol provides in Article 14 for the states' particular regard to the protection, preservation, restoration, and rational use of marine living resources. The Caspian states shall take all appropriate measures, based on the best scientific evidence available, to, first, develop and increase the potential of living resources for conservation, restoration, and rational use of environmental equilibrium in the course of satisfying human needs in nutrition and meeting social and economic objectives; second, maintain or restore populations of marine species at levels that can produce the maximum sustainable yield, as qualified by relevant environmental and economic factors and taking into consideration relationships among species; third, ensure that marine species are not endangered due to overexploitation; fourth, promote the development and use of selective fishing gear and practices that minimize waste in the catch of target species and that minimize the bycatch of nontarget species; fifth, protect, preserve, and restore endemic, rare, and endangered marine species; sixth, conserve biodiversity, habitats of rare and endangered species, as well as vulnerable ecosystems.

The overall legal context regarding the regime of the Caspian living resources was clarified in the framework of the multilateral state negotiations on the status of the Caspian Sea and finally defined in the Caspian Sea Convention adopted in 2018, as discussed below in detail.

7.4 Regulation of the Living Resources in the Caspian Sea According to the Caspian Sea Convention of 2018

The development of a clear regulation for the living resources in the Caspian Sea was necessary to enable riparian states to lawfully use the fish stocks as well as to protect them from overuse. The long-standing dispute over the extent of the coastal states'

---

[34]See: chapter XIX.

jurisdiction over the fish stock, as well as the territorial extension of the maritime zone in the Caspian Sea, was resolved only in 2018 with the signing of the Caspian Sea Convention.

7.4.1 Caspian Living Resources Regime in the Interstate Negotiations on the Convention on the Caspian Sea Legal Status

The Draft of the Caspian Sea Status Convention, which reflected the negotiation positions of the Caspian Sea countries, defined fishery zones as follows: *"Fishing zones are determined in accordance with this Convention and shall not extend further than [proposed 25–30 nautical miles, but not defined yet] nautical miles seawards from the baseline"* [Proposed by Azerbaijan, Iran, and Turkmenistan] *"or from the borderline of the Territorial Sea or the National jurisdiction zone"* [Proposed by Kazakhstan] (Article 9(10), Para. 1). *In its own fishing zone or in the zone of national jurisdiction [Proposed by Russia], each State Party exercises exclusive right on the conduct of the fishing industry and the usage of other living resources in accordance with relevant national legislation* (Article 9(10), Para. 2).

The definitions proposed in the draft of the Caspian Sea Convention did not clarify either the nature or the extent of the rights of the coastal states to the respective living resources in the Caspian Sea and their fishing zones.

According to the law of the sea, the provisions on the exclusive economic zone shall also extend to the fishing zones, recognizing coastal states' sovereign rights to the living natural resources therein. Russia, however, rejected such an understanding of the concept of fishing zones in the Caspian Sea. It rejected any legal forms that would ensure any sovereign rights in the Caspian Sea and supports the introduction of a zone of national jurisdiction, within which riparian states would possess merely certain nonresource-related rights. Such a position contributed to the legal uncertainty regarding the status of the areas where coastal states might, as proposed in the Draft Caspian Status Convention, exercise exclusive rights to fishing and the use of other living resources. The contentious point of the interstate negotiations on the legal status of the Caspian Sea was the extent of sovereignty that the coastal countries could exercise within the fishery zones.

In the previous decades of interstate negotiations, species like sturgeons and seals were discussed in the draft of the Caspian Status Convention, wherein the following solutions were proposed: *according to the present Convention and international mechanisms, the Parties shall jointly define the Total Allowable Catch of the valuable species of the Caspian living resource* (Article 9(10), Para. 3). *The Parties determine the capacity to harvest the living resources in their exclusive fishing zones or zone of national jurisdiction* (proposed by Russia). *If one of the Parties has no capacity to harvest the Total Allowable Catch, it may grant other States based on agreements or other arrangements and pursuant to the national legislation and regulations, access to the surplus of the Total Allowable Catch* (Article 9(10), Para. 4). *Natural and juridical persons of the Contracting States may, except the sturgeon and seals stocks, fish seawards of the exclusive fishing zones or the zones of national jurisdiction (proposed by Russia), according to the conditions laid down by the Contracting Parties in accordance with norms and rules of Para. 5 of this Article* (Article 9(10), Para. 5).

During the interstate negotiations after dissolution of the Soviet Union, the draft Caspian Status Convention allowed only the Caspian coastal states to participate in access to Total Allowable Catch (TAC) within the Fishery Zone of the Caspian Sea.

There were some regulations proposed concerning the quota system in the Caspian Sea, which were included into the final Draft of the Caspian Sea Convention. Russia proposed the following wording: *The sturgeon industry is prohibited, except within the mutually agreed by the Parties limited quotas of sturgeon catches in the Caspian Sea for the purpose of marine scientific research. The sturgeon industry close to Iran's coast is traditionally regulated by Iran, after consultation regarding its separate quotas* (Article 9(10), Para. 9). *The Parties determine the capacity to harvest the living resources in their exclusive fishing zones or zone of national jurisdiction* (proposed by Russia). *If one of the Parties has no capacity to harvest the Total Allowable Catch, it may grant other States based on agreements or other arrangements and pursuant to the national legislation and regulations, access to the surplus of the Total Allowable Catch* (Article 9(10), Para. 4). *The Parties shall jointly define the norms and rules in accordance with this Convention in particular related to: permits for fishing industry; catch quotas, fishing seasons and areas of fishing industry, the types, size and number of fishing gear; age and size of fish which may be caught; norms and rules* (Article 9(10), Para. 5).

7.4.2 *Regime for the Living Resources in the Caspian Sea Convention of 2018*

The Caspian Sea Convention defines and regulates the right and obligations of the Caspian Sea riparian states in respect of the use of the Caspian Sea, including its natural living resources (Article 2, Para. 2). Further on, the Caspian Sea Convention stipulates that its parties shall conduct activities in the Caspian Sea for the purpose of, among others, harvesting, using, and protecting the aquatic biological resources (Article 4). These activities shall be conducted in accordance with the Caspian Sea Convention, other agreements between the parties consistent with this Convention, and their national legislation. The Caspian Sea Convention defines aquatic biological resources as fish, shellfish, crustaceans, mammals, and other aquatic species of fauna and flora (Article 1). The Convention explains that harvesting is any type of activity aimed at removing aquatic biological resources from their natural habitat (Article 1). As 'shared aquatic biological resources' the Caspian Sea Convention defines aquatic biological resources, which are jointly managed by the parties (Article 1).

In terms of the principle according to which Caspian Sea states shall carry out their activities in the Caspian Sea, the Convention refers to the states' obligation relating to the

> application of agreed norms and rules related to the reproduction and regulation of the use of shared aquatic biological resources (Article 3, Para. 12).

The Caspian Sea Convention has introduced maritime zones where each of the coastal countries have different scope of sovereign rights on the living resources of the Caspian Sea. In Article 5, the Caspian Sea Convention provides that the water area of the Caspian Sea shall be divided into internal waters, territorial waters, fishery zones, and common maritime space. The first two—internal waters and territorial waters—fall under the territorial sovereignty of the coastal state, where the rights to living resources of those zones are fully subjected to the national authority of the relevant coastal state (Article 6).

The fishery zones and common maritime space introduced by the Convention are subject to the limited right of the coastal states. The fishery zone is defined

> as a belt of the sea where the coastal state holds an exclusive right to harvest aquatic biological resources (Article 1).

The Convention envisages 10-nautical-mile wideness for the fishery zones that are adjacent to the territorial waters of the Caspian Sea states (Article 9). The delimitation of the fishery zones has not been fully resolved but was left to be decided later on by the states with adjacent coasts, with due regard to the principles and norms of international law. The proposed concept of fishery zones refers to similar principles included in the law of the sea, with some important differences. The size of the indicated Caspian fishery zones has been defined for 10 nautical miles due to the small overall size of the Caspian Sea. The Caspian Sea Convention, with the establishment of the fishery zone, recognizes coastal states' sovereign rights to the natural resources living there. In the interstate negotiations over the legal status of the living resources, the legal regime of the fishery zone was a subject of dispute among the negotiating countries.

In view of the fact that the legitimacy of the valuable fishing industry with regard to preventing the degradation of living resources must be assessed in accordance with national legislation and international standards, to which not all coastal states are parties, this leads to a serious weakness of the Convention on the Status of the Caspian Sea in the regime of a fishing zone. The law of the sea guarantees coastal states' sovereign rights to the living resources within the EEZ, limited partly by the participation rights of third states.[35] The coastal states shall use the living resources[36] according the obligation to optimize the use of the resources.[37] There are no special

---

[35] Articles 62, 69, 70 UNCLOS.
[36] Articles 61, § 1, and 62 UNCLOS.
[37] See: (Hafner 1987), p. 272.

UNCLOS preconditions for assessing the living resources, but states shall consider certain factors[38] to ensure both the conservation of the living resources and the economic needs of the coastal states.[39] The conditions for access to the living resources are left to coastal states' national law regulations.[40] However, UNCLOS (Article 70) promotes particularly the rights of landlocked and geographically disadvantaged states.

The Caspian Sea Convention envisages special terms and procedures for the harvesting of shared aquatic biological resources, to be determined by a separate agreement to be concluded by all parties (Article 9, Para. 5).

> On the basis of this Convention and international mechanisms, the Parties shall jointly determine the Total Allowable Catch of shared aquatic biological resources in the Caspian Sea and divide it into national quotas (Article 9, Para. 3).

Ensuring states' exclusive national control over sturgeons and seal stocks is reasonable because of their economic importance and need of special protection. Setting quotas limiting the use of natural resources in the EEZ, respectively also in fishery zones, is common under the law of the sea. The usage rights used to be guaranteed exclusively to the respective coastal states and also certain private companies.[41] However, in case of the Caspian Sea Convention regulation, these rights are guaranteed exclusively to the Caspian coastal states.

According to the Caspian Sea Convention, the water area located outside the outher limits of the fishery zone is open to any type of uses by all the parties of the Convention (Article 1). It is so-called common maritime space, defined as one of the zones of the water areas of the Caspian Sea that remains under the joint sovereignty of the riparian states. The legal status of this zone is to some extent similar to the UNCLOS regulations on the so-called high sea zone. According to UNCLOS, the High Sea is characterized by the freedom of the sea and the freedom of fishing. The Caspian Sea Convention allows for any use of this area by the states, including the use its living resources, but limits the access to this zone merely to the Caspian Sea coastal states. This provision offers an important difference in comparison to the UNCLOS regulation of the High Seas' legal status. The Caspian Sea Convention offers here to its coastal states, exclusive rights overwaters of the Common Maritime Space, which may be classified as recognition of the joint sovereignty of these states over this area. This provision reflects rather the regime typical for regulation of water of international lakes which typically are used exclusively by states bordering these lakes. In opposite, UNCLOS guarantees the access to High Seas Zone for all states regardless of their geographical location and equipped them with all freedoms.

---

[38] Annotated Directory of inter-governmental organizations concerned with ocean affairs (A/CONF 62/L 14 from 10 Aug 1976).

[39] See: (Hafner 1987), pp. 267 et seq.

[40] Also: Article 297 § 3 (a) UNCLOS.

[41] See: (Hafner 1987), p. 119.

7.5 Conclusion

Extensive exploitation of the fish deposits of the Caspian Sea, whether conducted legally or illegally, as well as the industrialization of the whole Caspian region negatively contribute to the situation of the whole spectrum of living resources in the Caspian Sea. Protection guaranteed under the auspices of the CITES Convention could not yet stabilize the bio resource development not to exceed the maximum sustainable yield by the Caspian Coastal states. During the Soviet–Iranian period, fishery was under the regime of common use, excluding coastal fishery zones devoted to national use. Since the dissolution of the bilateral system of control over the Caspian Sea, the regime of the living resources has been regulated under the auspices of the International Commission on Aquatic Resources of the Caspian Sea (ICARCS), which is a regional and multilateral platform for Caspian state negotiations. However, it does not possess formal competence to adopt binding legal acts; it is, however, active in working out a legal framework for the use and protection of Caspian biological resources. The strengthening of the existing legal regime on the living resources was made through the conclusion of the Tehran Convention for the Protection of the Maritime Environment of the Caspian Sea. The recent adoption of an ancillary Protocol on Biological Diversity will strengthen the applicability of the Tehran Convention and the regime of protection of the Caspian living resources. It should, however, be well coordinated with the legislative activities undertaken under the auspices of ICARCS.

The framework offered by international maritime law, defining states' right of use of the living resources, as well as their obligation to protect them, found application in the newly adopted 2018 Caspian Sea Convention. Within the zones of the internal waters and territorial sea, the Caspian Sea Convention has equipped the states parties with full sovereignty over the living resources. The resources in the water area of the Caspian Sea beyond these two zones are called fishery zone and common maritime space. A fishery zone of 10 nautical miles has been defined as a belt of sea where a coastal state exercises an exclusive right to harvest aquatic biological resources defined as fish, shellfish, crustaceans, mammals, and other aquatic species of fauna and flora. The delimitation of these zones was left to the future agreements of the states with adjacent coasts. The Caspian Sea Convention provides that its parties shall jointly determine the total allowable catch of shared aquatic biological resources, which shall be jointly managed by the coastal states and divided into national quotas. The Caspian Sea Convention stipulates that if one of the countries is unable to harvest its entry quota in the total allowable catch, it shall grant access to the remaining quota to other coastal states of the Caspian Sea by way of an agreement.

According to the Caspian Sea Convention, a water area located outside the outer limits of the fishery zones is called common maritime space, which is open for use by all the parties. The legal status of the common maritime space reflects a concept of joint sovereignty, typical for the regulation of the waters of international lakes rather than of seas. In the law of the sea water area located outside the outer limits of the fishery zones are open to all countries regardless of their geographical location.

References

Anderson A (1975) Criteria for maximum economic yield of an international exploited fishery. In: Knight G (ed) The future of international management. West Publishers

Churchill R (1975) The fisheries jurisdiction cases: the contribution of the international court of justice to the debate on coastal states 'Fisheries Rights'. Int Comp Law Q 24

Garcia-Amador F (1974) The Latin American contribution to the development of the law of the sea. Am J Int Law 68:33–50

Hafner G (1987) Die seerechtliche Verteilung von Nutzungsrechten. Forschungen aus Staat und Recht, 71 ред. Springer, Heidelberg

Hey E (1987) The principle for the exploitation of transboundary marine fisheries resources. Kluwer, Dordrecht

Hyvarinen J, Wall E, Lutchman I (1998) United Nations and fisheries in 1998. Ocean Dev Int Law 29

Kindt J (1984) The law of the sea: Anadromous and Catadromous fish stocks, sedentary species and the highly migratory species. Syracuse J Int Law Com 11

Lagoni R (1992) Principle applicable to living resources occurring both within and without the exclusive Economic Zone or in Zones of Overlapping Claims. Report of the Committee of the International Law Association. Cairo, International Law Association Conference

Mamedov R (2001) International legal status of the Caspian Sea; issues of theory and practice. Turkish Yearb Int Relat 32:217–259

Ulfstein G (1998) The conflict between petroleum production, navigation and fisheries in international law. Ocean Dev Int Law 19

Zonn I, Zhiltsov S (2004) Kaspijskij Region (The Caspian region). Springer, Moscow

Chapter 8
The Legal Regime of the Pipelines in the Caspian Sea

8.1 Pipelines in the Caspian Sea

The oil and gas resources of the landlocked Caspian region are thousands of miles away from open sea ports, from where tankers could deliver them to markets in Europe, Asia, or America. This was one reason why a number of pipelines for the transport of liquids and gases (petroleum and natural gas from the Caspian fields) was built—namely, to transport resources overland for distances of several thousand kilometers. As some large oil and gas deposits are located seaward of the Caspian coast, there are plans to construct offshore pipelines on the seabed of the Caspian Sea.

Whereas oil and gas production in the Caspian Sea has been increasing from year to year, the expansion of export capacity has been slow. Many regional and global actors want to gain control of the Caspian's energy reserves and their transport routes to strengthen their own political presence in the region, to reduce their dependence on energy supplies from the Gulf region, or (as in the case of the new independent Caspian states) to secure their economic development. This complex geopolitical situation in the Caspian region impedes policy with respect to the laying of pipelines in the region.[1]

The two oil pipelines Baku–Novorossiysk (the northern route from 1997 and the second route from 2000) and Baku–Supsa transport oil from the fields of Azerbaijan to the west. The oil from Kazakhstan also flows through two lines: Atyrau to Samara in Russia, where it connects with the Russian main line, and since 2001 also through the pipeline of the Caspian Pipeline Consortium (CPC) from the oil field Tengiz to the Russian ports of Novorossiysk and Tuapse on the Black Sea. The required expansion of the loading capacity of the two ports, however, hampers further oil transport through the Black Sea to the Mediterranean area. An alternative to the CPC

---

[1] See: (Pradhan 2019), p. 474.

is offered by the oil pipeline Baku–Tbilisi–Ceyhan (BTC), which is designed for the transport of oil mainly from Azerbaijan to the world market.

The completion of other planned pipelines that could carry Caspian resources to the west and to the south and east in the near future is less foreseeable for political reasons. The strategic importance of the Caspian Sea natural resources can help in diversifying energy sources for many countries like China or the European Union states, which import their natural resources from the Middle East. But the new transportation routes require sufficient investment and political stability.[2] The strategy to deliver resources from the Caspian Sea to Europe is still high on the agenda. The European Commission used to promote the idea of a Trans-Caspian pipeline to guarantee the transport of gas from Turkmenistan through Azerbaijan and further on to the Southern Gas Corridor of Europe. The price of building the Trans-Caspian pipeline would be rather high and shall not be reasonably assumed to be covered by the European gas prices.[3]

However, the long political disagreement between the Caspian countries about building the Trans-Caspian pipeline seems to have been clarified in the Convention on the Legal Status of the Caspian Sea of 2018. However, there is still no clear interstate practice to assess whether the new legal framework will help to overcome the previous political constraints around the Trans-Caspian pipeline. As mentioned by Igor Bratchikov, Russian foreign ministry special envoy, interstate consultations regarding such pipeline "go on however long it takes," pointing to the endless talks, which might be interpreted as a Russian strategy in the Caspian Sea region concerning the trans-Caspian pipeline. According to the Protocol to the Framework Convention for the Protection of the Marine Environment of the Caspian Sea (2018), *when a contracting party has reasonable concerns that it would be affected by a significant transboundary impact of a proposed activity and when no notification has taken place in accordance with Para. 1 of this article, the party of origin shall, at the request of that contracting party, provides it with sufficient information whether a significant transboundary impact will take place as a result of the proposed activity* (Article 5, Para. 9). If necessary, these parties shall hold consultations regarding a possible participation in the environmental impact assessment procedure.[4] On the other hand, it was not Russia but another Caspian state that voted in favor of a more strict attitude toward pipeline construction in the Caspian Sea. During the work of the Preparatory Committee COP6 to the Tehran Convention,[5] it was Turkmenistan that suggested to take out the word "large diameter" from EIA Protocol Annex 1, which speaks about the construction of pipelines and the exploration on natural resources. Russia and Iran expressed their support to this proposed amendment, but Azerbaijan and Kazakhstan were against since such a change would apparently restrict the construction of all pipelines, large and small. Finally, the

---

[2]Pirani (2019).
[3]Ulviyye and Azhgaliyeva (2019). See also (Farah 2015)(n 4) 179–193.
[4]Kramer (2018).
[5]The Preparatory Committee for COP6, October 2017, 18.

protocol, adopted without any suggestions from Turkmenistan, provides in its Annex 1 for "large diameter pipelines." Still, Azerbaijan and Kazakhstan needed a long time to agree to the final wording of the EIA Protocol (see, e.g., COP4 2014 and the preparatory Committee for COP5 2017). It would be a future state practice to reveal what kind of future opportunities there will be for a lawful laying of submarine pipelines in and through the Caspian Sea.

The transport of Caspian natural resources to China is another important direction of export products from the Caspian countries. The Kazakh–China oil pipeline is China's first direct oil import pipeline from Central Asia, running from Kazakhstan's Caspian shore to Xinjiang in China. It was agreed upon in 1999, and its total length is 2800 km. China cooperation with Caspian states on energy resources is being conducted within a broader framework of the Belt and Road Initiative within which China invests in infrastructure including three gas pipelines from Turkmenistan to China. A large part of China's total gas import comes from Turkmenistan.[6] Emphasizing the great importance of Central Asia to China in the field of energy security, the Central Asia–China pipeline, starting in Turkmenistan, goes through Uzbekistan and Kazakhstan to Horgos, where it joins with China's West-to-East Gas pipeline. China's strategy to create an energy corridor allowing for a diversification of China's energy imports, next to the construction of CPIT pipeline (China–Pakistan–Iran–Turkey), foresees import of oil and gas by land routes from the Caspian Sea to reduce China's dependence on the import of these resources through Malacca Strait.

Turkmenistan's gas is fed also into the Turkmenistan–Afghanistan–Pakistan–India (TAPI) pipeline. But this project remains, however, economically infeasible due to sufficient length of the planned pipeline as well as security constraints along the planned routes of the pipeline. Its length is 1814 km, and it is supposed to serve for 30 years and was expected to be operational beginning 2020 (see footnote 8). The agreement on the construction of TAPI was signed in 2014; however, it would be more economically feasible to transport gas from the Caspian Sea rather than through Russia. Before 2009, the level of export from Turkmenistan to Russia was very high but later decreased due to political reasons.[7] Only in 2019 a new 5-years contract to supply of Turkmen gas to Russian was signed after the successful adherence of both states to the Caspian Sea Convention.

Given the geographical characteristics of the Caspian Sea as landlocked waters, the legal regime of lying pipelines to transport the natural resources has been subject to dispute between the coastal states. Much greater importance concerns the possible elaboration of legal practice defining the regime of pipelines on the seabed and underground of the Caspian Sea. The final set of rules determining the oil and gas pipeline regime shall reflecting international legal standards.

---

[6]Wang Yamei, China-Central Asia gas pipeline transports 47.9 billion cubic meters in 2019. http://www.xinhuanet.com/english/2020-01/06/c_138682150.htm accessed June 25, 2020.

[7]Petlevoy (2017) 6 see (Pirani and Henderson 2014; Henderson and Pirani 2015; Khatinoglu 2019).

8.2 International Law on Pipelines

International law provides detailed regulation of the regime for laying submarine pipelines. Pipelines located overland and at the bottom of the sea are a means of transporting petroleum and natural gas. The overland pipelines enjoy no special regime under international law. Submarine pipelines are regulated according to the law of the sea. The existing international rules concerning freedom of transport may apply to pipelines and gas lines when they are used for traffic in transit, if agreed upon by the contracting states concerned.[8] There are only a few treaties regulating this matter. A number of related issues (like property, licensing, safety standards, and the environment) are regulated by the national law of the individual states.

Most transboundary overland pipelines consist of separate parts located in areas falling under states' sovereignty, and their regime is therefore regulated by national laws. However, it is becoming increasingly common that transboundary pipelines are regulated by multilateral agreements.[9] They provide for the parties' general obligations regarding pipeline construction, nondiscrimination in usage, etc.[10] and can even give exact data on the delineation of the course for the laying of such pipelines.[11] The pipelines to be laid on the territory of another state for defense reasons require the permission of the state concerned.[12]

Initially, a state's freedom to lay submarine cables was recognized in the nineteenth century. This was regulated for the first time in the Convention for the Protection of Submarine Cables of 1884[13] and recognized as one of the freedoms of the high seas in 1927 by the Institute de Droit International.[14] The legal regime for laying and protecting submarine pipelines was set in the Convention on the High Seas of 1958 (Article 2(3)) and in UNCLOS (Article 87, Para. 1 (c)). The coastal state shall have the right to set conditions for pipelines entering its territory or territorial sea or to establish its jurisdiction over pipelines that are from other states and coming under its territory (Articles 79, 1–4 UNCLOS). Subject to its right to take reasonable measures for the exploration of the continental shelf, the exploitation of its natural resources, and the prevention, reduction, and control of pollution from pipelines, the coastal state may not impede the laying or maintenance of such cables or pipelines. The delineation of the course for the laying of such pipelines on the continental shelf is subject to the consent of the coastal state. The coastal state retains the right to establish conditions for pipelines entering its territory or territorial sea, as well as its jurisdiction over cables and pipelines constructed in or used in connection

---

[8] Art. 1 Para. of the Convention on Transit of Land-Locked States; Art. 124 Para. 2 UNCLOS.
[9] See: (Lagoni 1997), p. 1034.
[10] Brazil–Bolivia 1938, UNTS, vol. 51, p. 256; Brazil–Bolivia–Argentina–Paraguay–Uruguay, 1941 In: (Hudson and Sohn 1949/1950), vol. 8, p. 623.
[11] US–Canada, Northern Gas Pipeline Agreement, 1977.
[12] Haines–Fairbanks Oil Pipeline Agreement of 1955.
[13] See: (Martens 1817–1842), vol. 11, p. 281.
[14] Ann IDI, vol. 3 (1927), p. 339.

with the exploration of its continental shelf, along with the exploitation of its resources and the operations of artificial islands, installations, and structures within its jurisdiction.

Outside the territorial sea, states are free to lay submarine pipelines.[15] When laying submarine cables or pipelines, states shall have due regard to pipelines already in place.[16] In particular, the possibility to repair existing cables or pipelines shall not be prejudiced. In case of interruption or damage to a submarine pipeline by the owner of another submarine pipeline, the repair costs incurred by the pipeline owners must be borne by him.[17] The freedom to lay submarine pipelines— including the laying of new pipelines and repairing the old—as well as to enjoy other freedoms of the high seas shall be exercised by states with due regard to the interests of other states enjoying similar freedoms on the high seas and to those states' rights with respect to activities in the Area.[18]

A state's right to lay pipelines on the continental shelf and respectively on the seabed of the exclusive economic zone is limited by the following rights of the coastal state: to take reasonable measures for the exploration of the continental shelf, the exploitation of its natural resources, and the prevention, reduction, and control of pollution from pipelines;[19] also, the coastal state's consent is needed for the delineation of the course for the laying of such pipelines on the continental shelf (Article 79, Para. 3, UNCLOS). In the areas where the exclusive economic zone was established above the continental shelf, the legal regime of the continental shelf prevails. This applies except when the exclusive economic zone exceeds the seaward limits of the continental shelf as in such a territory the regime of the high seas is applicable (Article 58, Paras. 1, 2, UNCLOS).

The right of the coastal state to take reasonable measures for the prevention, reduction, and control of pollution from pipelines[20] includes its right to conduct an inspection and impose safety standards.[21] The pipes require pumping stations for their proper functioning. A safety zone will therefore be created around them.[22]

The maritime pipeline regime for the landlocked countries is a special case. Landlocked states are those countries that have no access to the seacoast. Their geographic location hampers their participation in world trade because they need to trade at a great distance from the sea, thus causing relatively high costs. Securing free

---

[15] Art. 2 (3), Art. 26, Para. 1, 1958 Convention on the High Seas; Art. 87 Para. 1 c), Art. 112 Para. 1 UNCLOS.

[16] See: (Martens 1817–1842), vol. 11, p. 281.

[17] Art. 4 Convention for the Protection of Submarine Cables; Art. 28 1958 Convention on the High Seas, 1958; Art. 114 UNCLOS.

[18] Art. 2 1958 Convention on the High Seas, 1958; Art. 87 Para. 2, Art. 150, 153 UNCLOS.

[19] Art. 4 Convention on the Continental Shelf 1958 Art. 26 Para. 1, 2 1958 Convention on the High Seas, Art. 79 Abs. 2 UNCLOS.

[20] Art. 4 Convention on the Continental Shelf; Art. 26 Abs. 1, 2 1958 Convention on the High Seas; Art. 79 Abs. 1, 2 UNCLOS.

[21] Art. 27–29 1958 Convention on the High Seas; Art. 113–115 UNCLOS.

[22] Art. 5 Convention on the Continental Shelf; Art. 60, Art. 80 Abs. 4–7 UNCLOS.

and unfettered access to the high seas is of great significance for landlocked countries, which in turn is connected with transit issues. Both persons and property originating from a landlocked state or arriving at its territory must cross the territory of another state, which matter can cause numerous legal, political, and administrative difficulties.[23]

Not all international agreements that guarantee freedom of transit extend to the rights of landlocked countries to lay pipelines. The Barcelona Convention and the Barcelona Statute on the Freedom of Transit, the first legal source that has provided for the freedom of transit, does not apply to laying of pipelines. The Barcelona Convention and the Statute have a more general scope of application in comparison with the New York Convention on Transit Trade of Landlocked States of 1965 (further referred to as the New York Convention of 1965). The latest, alongside traditional, means of transport includes rules for "other" means of transport, including oil and gas pipelines, which shall be established by a common agreement between the contracting states concerned, with due regard to the multilateral international conventions to which these states are parties.[24] UNCLOS, while defining the means of transport, states that landlocked states and transit states may, by agreement between them, include as means of transport pipelines and gas lines (Article 124, Para. 2). Also, the General Agreement on Tariffs and Trade (GATT) secures the transit right of the landlocked states (Article V), which may only be exercised by states and not private enterprises (Article XVII). None of these provisions allowing for pipeline transit rights may be applicable to the Caspian Sea pipeline because neither the Soviet Union itself nor its successor states have ever become parties to these conventions. Therefore, a particularly important role for the expansion of the pipeline transit rights of Caspian landlocked countries is played by the Energy Charter, which was signed in Lisbon on December 17, 1994, by all the states of the former Soviet Union. Its weakness lies in the fact that Russia has not ratified the Charter.[25]

The Energy Charter defines (Article 1, Para. 4, 5) oil transportation as one of the priority areas for regulation. Its provisions are applicable to all economic activities in the energy sector, including the transportation of primary energy sources (oil and gas) and energy products. It obliges parties to take necessary measures to facilitate the transit of energy materials and products consistent with the principle of freedom of transit and without distinction as to the origin, destination, or ownership of such energy materials and products or discrimination as to pricing based on such distinctions and without imposing any unreasonable delays, restrictions, or charges (Article 7.1). Contracting parties shall encourage relevant entities to cooperate in, first, modernizing energy transport facilities necessary for the transit of energy materials

---

[23] See: (Uprety 1995).

[24] Art. 2 Para. 1 New York Convention of 1965.

[25] After signed the Energy Charter Treaty in 1994 Russia accepted its provisional application (agreeing to apply its provisions as far as they are with its national law), which was terminated by Russia in 2009.

and products; second, developing and operating energy transport facilities serving the Areas of more than one contracting party; third, applying measures to mitigate the effects of interruptions in the supply of energy materials and products; and, fourth, facilitating the interconnection of energy transport facilities (Article 7, Para. 2). The next, very important provision states that in the event that the transit of energy materials and products cannot be achieved on commercial terms by means of energy transport facilities, the contracting parties shall not place obstacles in the way of new capacity being established, except as may otherwise be provided in applicable legislation regarding environmental protection, land use, safety, or technical standards. In the event that the transit of energy materials and products cannot be achieved on commercial terms by means of energy transport facilities, the contracting parties shall not place obstacles in the way of new capacity being established (Article 7.4). However, a party through whose territory primary energy sources and energy products can be routed in transit is not obliged to permit the construction or modification of energy transport facilities or a new or additional transit through existing energy transport facilities in case it would endanger the security or efficiency of its energy systems, including the security of supply (Article 7, Para. 5). A special system of dispute settlement described in the Charter may be applicable only following the exhaustion of all relevant contractual or other dispute resolution remedies previously agreed upon between the contracting parties involved in the dispute (Article 7, Para. 7).

8.3 Regulations on Pipelines in the Caspian Sea as Reflected in the Multilateral Negotiations Among the Caspian Sea States Until 2018

The pipeline regime in the Caspian Sea has never been subject to separate interstate regulation. Within the intergovernmental negotiations reflected in the draft of the Caspian Sea Convention discussed in the 1990s, there was an attempt to regulate the pipeline regime. The Draft Caspian Status Convention was—albeit indirectly—following law of the sea provisions. *Contracting states may lay submarine cables and pipelines on the bottom of the Caspian Sea [proposed by Azerbaijan, Kazakhstan and Turkmenistan] in accordance with this Convention, international legal standards and agreed economic standards* (Article 13(2), Section 1) (proposed by Iran); *The delineation of the course for the laying of such pipelines is subject to the consent of the state party, if the submarine pipe is to be laid through the mining site of the coastal state* (Article 13(2), Para. 2) (proposed by Azerbaijan, Kazakhstan, and Turkmenistan);

Nothing affects the right of the state parties to establish conditions for laying pipelines entering their mining sites on the seabed (Article 13(2), Para. 3) (proposed by Azerbaijan, Kazakhstan, Iran, and Turkmenistan).

Regarding the laying of a trans-Caspian pipeline, the states parties differed seriously in their positions. Provisions proposed by Iran, together with Russia, were not compatible with existing international law norms regarding the laying of submarine pipelines. *Contracting states may lay submarine cables and pipelines on the bottom of the Caspian Sea* (Article 13(2), Para. 1) (proposed by Russia and Iran); *States Parties establish conditions for the laying of technological pipelines in their own sectors or their zones at the seabed of the Caspian Sea* (Article 13(2), Para. 2) (proposed by Russia and Iran); *The Contracting states may lay submarine main pipelines on the floor of the Caspian Sea, under the condition that an ecological expertise of these projects will be approved by all the coastal countries. The state laying the pipeline shall bear material responsibility for damages caused to the other Parties and to the marine environment occurring due to break up of the pipeline* (Article 13(2), Para. 3) (proposed by Russia and Iran).

Iran and Russia ruled out the possibility of a unilateral decision with regard to the laying of a Trans-Caspian pipeline. Russia's position on the division of the Caspian Sea into zones of national jurisdiction, where coastal states' rights have no sovereign character, excluded coastal states' privileges recognized by UNCLOS regarding rights on laying pipelines. Azerbaijan, Kazakhstan, and Turkmenistan represent the position that each of the coastal states exercises the right to lay a submarine Trans-Caspian pipeline. Such a right has been based on an agreement concluded exclusively between states whose seabed mining site is crossed by the routes of pipelines. This proposal would apply with the provisions of UNCLOS relating to the rights and obligations of states on the laying of submarine pipelines. It required the consent of the respective coastal state (Article 79, Para. 3) because nothing must affect the right of the coastal state to establish conditions for pipelines entering its territory or its jurisdiction over pipelines constructed.

8.4 Regulations of Pipelines in the Caspian Sea as Reflected in the Caspian Sea Convention of 2018

Perhaps one of the greatest innovation in the Caspian Sea Convention was the establishment of the legal framework for the construction of submarine pipelines in the Caspian Sea. The Caspian Sea Convention, which regulates the delimitation of the marine zones in the Caspian Sea, has clarified respectively the regime of the laying of pipelines, as well acknowledged the rights of the coastal states to build Trans-Caspian pipelines. According to the Caspian Sea Convention:

> Parties may lay submarine pipelines on the bed of the Caspian Sea (Article 14, Para. 1).

This provision was one of the most contested during the whole period of negotiations over the legal status of the Caspian Sea; however, it has been successfully settled in the new Caspian Sea Convention. This provision reflects the right of the states to lay submarine pipelines in the territories beyond the territorial sea, which in the Law of the Sea has been recognized in the regulations of the continental shelf and Area (Article 79, Para. 1, UNCLOS), which can in the future have significance for the transport of Caspian Sea resources to European countries.[26]

An important limitation to the laying of submarine pipelines on the seabed of the Caspian Sea was, however, imposed by the Caspian Sea Convention:

> The Parties may lay trunk submarine pipelines on the bed of the Caspian Sea, on the condition that their projects comply with environmental standards and requirements embodied in the international agreements to which they are parties, including the Framework Convention for the Protection of the Marine Environment of the Caspian Sea and its relevant protocols (Article 14, Para. 2).

The parties may lay trunk submarine pipelines on the bed of the Caspian Sea, on the condition that their projects comply with environmental standards and the requirements embodied in the international agreements to which they are parties. One of such agreements is the Framework Convention for the Protection of the Marine Environment of the Caspian Sea, otherwise known as the Tehran Convention. Relevant protocols from the said Convention include the Aktau Protocol on Preparedness, Response, and Cooperation for the Protection of the Caspian Sea from the oil pollution caused by seabed activities, which will be discussed in more detail. Another important requirement for laying the submarine pipelines on the Caspian Sea seabed is the one that states that the routes of pipelines shall be determined by an agreement with the party the seabed sector of which is to be crossed by the pipeline (Article 14, Para. 3). It is a provision that is similar to the conditions included in UNCLOS (Article 79, Para. 3, UNCLOS).

8.5 Conclusion

Although the oil and gas resources of the Caspian Sea are vast, their transportation to world markets requires enhancement. The most suitable form for doing so would be to ship the resources over a waterway, which is limited in the case of the Caspian region, or to use pipelines. The existing network of pipelines from the Caspian Sea shall be extended in accordance with international legal standards. For the land pipelines, there are different legal frameworks available, such as the Energy Charter of 1994, where the newly independent riparian states are party to. In case maritime pipelines are to pass the bottom of the sea, the law of the sea would be respectively applicable.

In the past, the legal regime of the Caspian maritime pipelines was never subjected to interstate agreements. It was subordinated only to the general practice of the Caspian states in regulating the use of the Caspian Sea. Nowadays, according to the Convention on the Caspian Status of 2018, submarine pipelines on the seabed in the Caspian Sea are lawful. They must respect the rights of the coastal states, which exercise their sovereign rights in the sectors over the whole bottom and

---

[26]Ulviyye and Azhgaliyeva (2019).

subsoil of the Caspian Sea beyond the so-called territorial waters. The routes of submarine pipelines shall be determined by an agreement with the party the seabed sector of which is to be crossed by the pipeline. They should comply also with environmental standards embodied in the international agreements to which the Caspian Sea states are parties. Among them, the most important would be the Framework Convention on the Protection on the Maritime Environment of the Caspian Sea (so-called Tehran Convention and its additional protocols). The Caspian Sea Convention has acknowledged the rights of the coastal states and to build a Trans-Caspian pipeline, which was most contested in the period of negotiations over the legal status of the Caspian Sea. This provision reflects the regulation of UNCLOS on the continental shelf. There is, however, no clear state practice yet to see how the provisions will be implemented in the future.

References

Farah P (2015) Energy security, water resources and economic development in Central Asia. In: Rossi P, Farah PD (eds) Energy: policy, legal and social-economic issues under the dimensions of sustainability and security, World Scientific Reference on Globalisation in Eurasia and the Pacific Rim. Imperial College Press & World Scientific Publishing, London, pp 179–193

Henderson J, Pirani S (2015) The Russian gas matrix: how markets are driving change. Energy J 35 (4):357–359

Khatinoglu D (2019) Turkmenistan resumes flows to Russia. Azerbaijan: www.naturalgasworld.com

Kramer AE (2018) In a Prize for Big Oil Firms, Caspian Deal Eases Access. [Online] Available at: https://www.nytimes.com/2018/10/08/business/energy-environment/caspian-sea-deal-eases-access.html. Accessed 20 June 2020

Lagoni R (1997) Cables, submarine. In: Bernhardt R (ed) Encyclopaedia of public international law, 3rd edn. Elsevier

Martens G (1817–1842) Nouveau recueil de traités ... depuis 1808 jusqu'à présent, 7th edn. Nouveau recueil de traités depuis 1808 jusqu'à présent. Forgotten Books, Göttingen

Petlevoy V (2017) Gazprom zakliuchil piatiletnyi kontrakt s Uzbekistanom (Gazprom signed a five-year contract with Uzbekistan). Vedomosti.ru, Moscow

Pirani S (2019) Central Asian Gas: prospects for the 2020s. Oxford Institute for Energy Studies

Pirani S, Henderson J (2014) Central Asian and Caspian gas for Russia's balance. Oxford Institute for Energy Studies, 155

Pradhan R (2019) Petropolitics and pipeline diplomacy in Central Asia: can India afford to wait in the wings? India Q J Int Aff 75(4):472–489

Ulviyye A, Azhgaliyeva D (2019) Assessing energy security In: The Caspian region: the geopolitical implications for European strategy, 1011 edn. s.l.: ADBI Working Paper Series

Uprety K (1995) Right of access to the sea of land-locked states: retrospect and prospect for development. J Int Leg Stud 21 (winter)

Chapter 9
The Legal Regime of Maritime Navigation on the Caspian Sea

9.1 Ship Navigation on the Caspian Sea

Shipping has always been one of the most important means for developing international trade and economic relations between Caspian countries. The Caspian Sea covers an area of about 371,000² km and has no natural connection with the oceans. However, via the Volga river, the Volga–Don Canal, and across the Don, there is a navigable linkage to the Sea of Azov and thus to the Black Sea, the Mediterranean, and the Atlantic. The Volga waterway can facilitate maritime traffic from the Caspian Sea through the Volga–Baltic Canal to the Baltic Sea. The main ports on the Caspian Sea are Astrakhan, Olya, Makhachkala (Russia), Baku (Azerbaijan), Aktau (Kazakhstan), Bandar-e Anzali (Iran), and Turkmenbashi, formerly Krasnovodsk (Turkmenistan). At the time of Tsar Peter the Great, Russia sought to establish itself as the main maritime power in the Caspian Sea. Thanks to his success in the Russian–Persian War, Russia became the ruling power in the whole region. From the October Revolution to the beginning of 1990, because of the isolation policy of the Soviet Union, the route through the Caspian Sea and the Volga was closed for international transportation.[1] Based on agreements between the USSR and Iran, all ships not flying the flag of the USSR or Iran were excluded from operating in the Caspian Sea. After the collapse of the Soviet Union, the Caspian Sea gained major geopolitical and economic importance and now facilitates international trade between 26 Asian and European states.

The most important trade routes were established as early as the beginning of the eighteenth century. Russian ships sailed regularly from Astrakhan, on the eastern coast of the Caspian Sea, for Kabak harbor and Karagan. There were also trade routes between Astrakhan and Baku, Derbent and Nisabad.[2] Shipping transport was revived when in 2000 an agreement was signed between the Russian Federation,

---
[1] See: (Arsenov 2003), p. 8.
[2] See: (Tuschin 1978).

India, and Iran on the creation of the international transport corridor "North–south." This corridor was designed to bring goods from India, Pakistan, and the Persian Gulf through the territory of Iran and its harbors on the Caspian Sea and then further on through Russia's ports to the countries of Central and Eastern Europe and Scandinavia. The Astrakhan and Olya traffic nodes, two of the key elements of this traffic corridor, have recently gained in significance. Since 2002, there has also been a ferry route—namely, the "Caspian Tracker Line"—for shipping goods between this port and two other Caspian ports (Anzali in Iran and Aktau in Kazakhstan). Another important element of this route is the Russian harbor of Makhachkala. Iran competes with Russia through large international free trade zones established in the city of Anzali. The "North–south" Transport Corridor is also of great importance for Kazakhstan, which ships its oil resources via the Aktau harbor. In Soviet times, there was only one ferry line that crossed the Caspian Sea from Baku to Krasnovodsk (now Turkmenbashi). It was established in 1929 and operates regularly to this day alongside the ferry line of Makhachkala–Turkmenbashi and Aktau–Baku. Today, there are four ferry terminals: in Baku, Makhachkala, Turkmenbashi, and Aktau.

Russia's policy is to create corridors for sustainable transport, and in 2017 Russia adopted a strategy for the development of Caspian Sea seaports till 2030. Its goal is to develop tourism and increase trade with Iran, India, and Gulf states via the extendable Makhachkala, Astrakhan, Olya, and Kaspiysk ports. On the other hand Russia aims an expending Eurasia Channel between the Caspian Sea and Black Sea to enhance trade through Caspian Region towards China.[3]

The Transport Corridor Europe Caucasus Asia (TRACECA) program was launched by the European Union in 1993 as a new corridor linking Europe to Central Asia across the Black Sea, the Caucasus, and the Caspian Sea. TRACECA offered EU-infrastructure-related assistance but recently seems to have lost momentum.[4] A new route through the Caspian Sea is one of corridors of the Belt and Road initiative, which was launched in 2013.[5] China and Europe are connected by the Trans-Caspian International Transport Route, which has many advantages for China. A Protocol on a Trans-Caspian International Transport was signed in 2017 by Kazakhstan, Azerbaijan, and Georgia. Unlike other corridors, the Piraeus–Khorgos route for the most part passes through water and thus forms the smallest number of intermediary countries (including Georgia, Azerbaijan, and Kazakhstan). It also bypasses Russia,

---

[3]Potential for Development of Caspian States Cooperation https://moderndiplomacy.eu/2019/12/18/potential-for-development-of-caspian-states-cooperation/ accessed July 15, 2020.

[4]TRASECA website. http://www.traceca-org.org/kz/home/ accessed July 15, 2020.

[5]"Kaspiiskii Vestmik": Kaspiiskii region v Ekonomicheskom poiase Shelkovogo puti: kitaiskii vzgliad (Septeber 23rd, 2018). http://casp-geo.ru/kaspijskij-region-v-ekonomicheskom-poyase-shelkovogo-puti-epshp-kitajskij-vzglyad/ accessed July 15, 2020.

Alash: Kaspiiskoe more – luchshii vybor Kitaia dlia Odnogo poiasa i puti. See: https://bit.ly/2VK5NO4 accessed July 15, 2020.

One Belt, One Road (OBOR):China's regional integration initiative (July, 2016) https://www.europarl.europa.eu/RegData/etudes/BRIE/2016/586608/EPRS_BRI(2016)586608_EN.pdf accessed July 15, 2020.

which helps to strengthen the cooperation between China and the Caspian littoral states. The Trans-Caspian International Transport Route connects the Greek port of Athens with the Georgian terminals in Batumi and Anaklia on the Black Sea.

The growing interest of states in shipping requires the establishment of a new legal regime for the Caspian Sea, which would ensure the application of internationally binding legal norms to Caspian shipping.

9.2 The Legal Regime of Shipping in International Law

There is no binding, overarching international legal definition of what is meant by the notion of ship or shipping. Maritime shipping differs from inland navigation, especially in terms of the spatial area of shipping operations. Maritime vessels operate mainly in areas outside of national jurisdiction. Maritime shipping is regulated, among others, by the international law of the sea since maritime vessels operate mainly in areas outside of national jurisdiction. International agreements, which regulate maritime shipping, include two different sets of regulation. The first group, which belongs to public international law, regulates the duties and rights among riparian states. The other set of provisions regulates the relations between subjects of private law, which is not the focus of this book.[6]

The United Nations Convention on the Law of the Sea (UNCLOS) defines separate legal regimes for navigation in the different maritime zones. The legal regime of internal waters is similar to the regime of the land territory in terms of navigation and passage. There is no right of innocent passage in the internal waters. In the next adjacent zone, known as the territorial sea, vessels under the flag of the coastal state, as well as foreign vessels, can exploit the right of navigation and enjoy the right of innocent passage through the territorial sea (Article 17, UNCLOS). In the exclusive economic zone (EEZ), the regime of the limited freedom of navigation applies accordingly to UNCLOS Article 58, Para. 1, which can be exercised by the subjects of the relevant provisions of this Convention, including the coastal states, which exercise sovereign rights over the resources in their EEZ and other rights related to artificial islands, maritime protection, etc. Beyond the EEZ, the principle of the freedom of the high sea prevails.[7] It includes the freedom of navigation to be exercised by all states with due regard to the interests of other states, an indispensable prerequisite for international trade, free of interference by third states. The principle of the freedom of navigation does not cover commerce-related questions but merely defines the scope of the free movement of ships through the maritime zones. The state practice in the implementation of the legal principle of freedom of the sea and freedom of navigation proves that they both shall not be understood in a wide economic sense but are separated from the guarantees of cargo and harbor

---

[6]Bruce (1993).
[7]Ipsen (2004), p. 289.

access, nondiscriminatory treatment, and claims to participate in international maritime trade on an equitable basis without discrimination.[8]

The rights of the coastal states over their nationals, vessels, and installations and structures in the Caspian Sea reflect the legal principles for the territorial sea and the exclusive economic zone as regulated in the UN Convention on Law of the Sea (UNCLOS) of 1982. UNCLOS provides no rights to the foreign merchant ships or warships, except in cases of emergency, or according to other international agreements, to either enter or call at national port facilities, which is an imminent part of national state territory.[9] In the contiguous zone, the coastal state has no special right to navigation but may only exercise the control necessary to prevent or punish infringements of its customs, fiscal, immigration, or sanitary laws and regulations within its territory or territorial sea. In the exclusive economic zone (also in the fishing zone), all states, whether coastal or landlocked, enjoy freedom of navigation, as prescribed for the high sea (Article 58, UNCLOS). The only restriction for the freedom of navigation reflects the coastal states' sovereign right to the protection and preservation of the marine environment (Article 56 b) iii), UNCLOS), especially their legislative powers against pollution caused by ships (Articles 210, 211, 216, 220, UNCLOS). According to UNCLOS, the freedom of navigation[10] constitutes one of the main "maritime freedoms," and it serves as a basis for the principle of freedom of the high seas. Thus, within the freedom of each state, whether coastal or landlocked, is the right to sail ships flying its flag on the high seas.[11]

The law of the sea provides for the right of innocent passage for merchant ships, and the rules governing the passage of warships and other government ships operated for noncommercial purposes are not clear enough. The Geneva Convention on the Territorial Sea and the Contiguous Zone of 1958 and UNCLOS can be interpreted as an approval for the innocent passage of warships. However, a number of states expressed serious reservations about the fact that the states retain the right to control or approve the peaceful passage of warships.[12] The requirement for the earlier approval by the coastal state or at least the duty of announcing the passage was not accepted during the Third Law of the Sea Conference.[13] This condition did not become part of common law either.[14]

The numerous international agreements include the principle of national treatment, where parties wish to ensure that their nationals, goods, ships, etc. are treated

---

[8]Lagoni (1988).

[9]In case of Nicaragua v. United States, ICJ Rep. 1986, 111.

[10]Freedom and Safety of Navigation out of coastal sea area is regulated by UNCLOS (Part XII, Section 7) and other Conventions like Convention on the International Regulations for Preventing Collisions at Sea and other rules set by International Maritime Organisation, which role is to impartially monitor navigational routs.

[11]Article. 90 et seq. UNCLOS.

[12]See: (Kasoulides 1992), p. 146.

[13]Article 21 UN Document A/AC. 138/SC. II/L.18.

[14]See: (Vukas 2004), p. 141.

equally by the other contracting states.[15] This clause means a ban on discrimination against national ships. It confirms that all benefits granted to a merchant ship flying the flag of a contracting state must be granted also to a merchant ship flying flags of all other states parties. Thus, foreign goods and services and their providers must not be treated less favorably than domestic ones. In international legal practice, the national treatment principle, whose origin lies in the principle of freedom of transit,[16] is very common, for instance with respect to landlocked countries[17] or in the General Agreement on Tariffs and Trade (GATT).[18]

Ships shall sail under the flag of one state only and shall be subject to its exclusive jurisdiction on the high seas (Article 92 et seq., UNCLOS), which covers administrative, technical, and social matters of ships. Although a coastal state has full territorial sovereignty over the territorial sea, criminal and civil jurisdiction can be exercised by the coastal state on board a foreign ship passing through the territorial sea but merely in limited cases (Articles 27, 28, UNCLOS). The general principle of international law is that the nationals of a state are exclusively subject to its jurisdiction. Neither a warship, which encounters on the High Seas nor foreign ship, other than a ship entitled to complete immunity, is justified in boarding it unless there are reasonable grounds for suspecting that the ship is engaged in piracy, in the slave trade, etc. (Article 110, UNCLOS). Hot pursuit of a foreign ship may be undertaken when the competent authorities of the coastal state have good reason to believe that the ship has violated the laws and regulations of that state (Article 111, UNCLOS).

UNCLOS contains standards for the international management of safety zones, regulated in the framework of the exclusive economic zone, where due notice must be given of the construction of artificial islands, installations, or structures, and permanent means of giving warning of their presence must be maintained (Article 60, Para. 3). In the case of the necessity to protect artificial islands, installations, or structures, the coastal state may establish an adequate safety zone. Within the EEZ, the state may take appropriate measures to ensure the safety of navigation and of artificial islands, installations, and structures. The breadth of the safety zone is determined by the coastal state, by considering applicable international standards. These zones must be designed so that they take into consideration the nature and function of artificial islands, installations, or structures. They shall not extend over a distance of 500 m around the safety zones, measured from each point of the outer edge of the artificial islands, installations, or structures.

The development of international shipping is stimulated by economic needs. The local circumstances, traditions, experience, and developmental stage of countries' shipping, as well as their politics and economics, are usually very diverse. For setting

---

[15] See: (Fox 1992), p. 296.

[16] Article 2 Convention on Freedom of Transit of 1921; Article 5 GATT.

[17] See: (Uprety 1995); Article 3 Abs. 1 Convention on the High Seas of 1958; Article 15 Convention on Transit of Land-Locked States, 1965.

[18] Articles 3, 17 GATT Convention.

up a successful shipping regime, two conditions are to be met: first, the guarantee of freedom of trade and services and, second, the existence of an efficient information system that facilitates and enables trade.[19]

The international agreements that deal with maritime shipping are of a different legal nature. The first group regulates the duties and rights among states, and the other regulates the relations between subjects of private law.[20] The law of the sea sets out legal standards such as the right of passage through international channels, waterways, and straits, as well as the legal status of ships in harbors and the rights of landlocked states. Maritime law was usually adopted at the initiative of actors of private law, which negotiated with states via the Committee Maritime International (CMI). This practice ended, however, when it was replaced by the International Maritime Organization (IMO), until 1982 known as the Inter-Governmental Maritime Consultative Organization (IMCO).

Public-sector involvement in defining the technical standards of shipping, its safety and stability, routes, social problems of the crew, pollution control, etc. has increased significantly in recent years. Not only will new binding standards be introduced, but so will penalties for law violations. This development is determined by the constant changes in the political relations among states, which bring significant instability to worldwide shipping. An example is the rejection of shipping rules by the Soviet Union after the Cuban Missile Crisis of 1962, which remained out of use despite diplomatic efforts during conferences in Moscow and Hamburg at the beginning of the 1980s.

Such a deplorable state of affairs was created by the years of distrust that prevailed from the late 1970s to the early 1980s between the United States, the European countries, and the Consultative Shipping Group. A positive development was visible merely in the international shipping industry of the Western European countries, which gradually integrated maritime trade laws enhancing free shipping zones.[21] An important normative contribution to the development of international shipping was also provided by international organizations. For example, the Organisation for Economic Co-operation and Development (OECD) adopted in the 1980s 13 basic principles of maritime transport policy. Also, the United Nations Conference on Trade and Development (UNCTAD) and its Shipping Committee drew up the "UNCTAD Liner Code."[22] This Code includes basic principles of standard commercial practice and aims to become a "universally acceptable code for liner conferences." The Code never came into force but did become the basis for subsequent legislative codifications in this field.[23]

---

[19] See: (Goss 1985), pp. 391 et seq.

[20] See: (Farthing and Brownrigg 1993), p. 33.

[21] First was the Regulation 4055/86 applying the principle of freedom to provide services to maritime transport between Member States and between Member States and third countries.

[22] Convention on a Code of Conduct for Liner Conferences 1974.

[23] EC Regulation 954/79; United Nations Convention on conditions for registration of ships 1986.

9.2 The Legal Regime of Shipping in International Law

An important area of legal regulation is safety at sea. By "safety at sea," we are to understand the safety of shipping and the safety of life and goods, as well as the safety of the maritime environment.[24] The issue of the vessel's safety refers to the construction of the vessel and its classification, its equipment, as well as the nature and operation of its load. In the case of passenger ships, additional attention is paid to the measures for safeguarding life, and they are to be guaranteed by special equipment. A maritime safety system is to be defined as certain standards and conditions that include a system of legal standards regarding safety at sea,[25] a system of institutions responsible for setting legal standards for maritime safety,[26] a system of institutions enforcing the norms of safety at sea,[27] and a system of institutions responsible for updating and spreading information on shipping.[28] The high importance of the issue of shipping safely exceeds the issue of cargo and passengers and also covers all other human activities at sea, such as fishing, marine scientific research, exploration and exploitation of the seabed, and environmental protection. The international safety standards are of a diverse character. A distinction is made between, firstly, technical standards that shall reduce the risk of accidents or the possible consequences of an accident; secondly, protection standards for the environmental-friendly use of ships, with the exception of accident situations; thirdly, construction standards for ships and port construction; fourthly, qualification standards for the crew of the ships.[29]

The current shipping standards, recognized worldwide, are mainly included in the International Convention of 1974 for the Safety of Life at Sea (SOLAS) and other international agreements.[30] One of the most important steps contributing to the enhancement of safety at sea initiatives was the establishment of the Society of Lloyd's Register in 1834, which contains detailed information on all vessels. One of the most important features of navigation is its freedom to pass without any obstacles both on the high seas and in the territorial waters of other countries. Freedom of navigation is merrily limited by provisions aimed at prevention of collisions at sea. This task of ensuring the safety of ships is carried out by the International Maritime Organization, originally known as Inter-Governmental Maritime Consultative

---

[24]See: (Łukaszuk 1997), p. 104.

[25]UNCLOS; Convention on the International Regulations for Preventing Collisions at Sea, 1972 (COLREG); SOLAS Convention, 1974; SAR Convention, 1979; MARPOL.

[26]International Maritime Organisation (IMO); International Hydrographic Organization (IHO); International Association of Marine Aids to Navigation and Lighthouse Authorities (IALA); Standardisation (ISO); Committee' International Radio-Maritime (CIRM); World Meteorological Organization (WMO).

[27]Sea chambers, Civil and Criminal Courts, Administrative authority of coastal state, Insurance company.

[28]Global Maritime Distress and Safety System (GMDSS); Maritime Safety Broadcasts (Navtex); Organisation on Maritime Search and Rescue (SAR).

[29]See: (Łukaszuk 1997), p. 116.

[30]1966 Load Line Convention; MARPOL; International Bulk Chemical Code, International Bulk Chemical Code; International Carrier Code.

Organization. All respective standards relating to maritime safety were defined in the International Regulations for Preventing Collisions at Sea 1972 (Colregs) and related agreements.

A sustained regulation of shipping requires the drawing up of certain rules related to the environmental protection of the sea. The first International Convention for the prevention of pollution of the sea by oil was adopted in 1954. A significant adjustment to the international protection standards came only after 1967, when the huge Torrey Canyon oil tanker sunk. In the wake of this disaster, the International Convention for the Prevention of Marine Pollution from Ships (MARPOL) was adopted, along with some additional agreements.

9.3 Historical Rules on Navigation in the Caspian Sea

The international legal standards for navigation are not directly reflected in the existing agreements regulating navigation in the Caspian Sea. Until the end of the twentieth century, the rights on navigation in the Caspian Sea were exercised exclusively by Iran and Russia. Russia received access to the Caspian Sea not until the eighteenth century. That is when the first agreements between Russia and Persia were concluded, which is to be considered as the beginning of the formulation of the international legal status of the Caspian Sea as they were devoted to the regulation of navigation. In the Treaty of St. Petersburg of 12 September 1723, Persia recognized Russia's exclusive navigation rights in the Caspian Sea for a period of 10 years. After losing the wars during the nineteenth century, Persia's navigational rights in the Caspian Sea were further limited by the Treaty of Gulistan of 1813 and the Treaty of Turkmenchay of 1828. Their merchant ships neither received rights to shipping in the Caspian Sea: only Russia's trade vessels were eligible for navigation there. This was not lifted until the conclusion of the Treaty of Friendship between the USSR and Persia in 1921. This agreement, which remains in force until today, confirmed the unlimited freedom of navigation in the entire Caspian Sea for ships flying the flag of one of the coastal states.[31] This freedom of navigation in the Caspian region, which was confirmed by subsequent treaties concluded between the USSR and Iran in 1931, 1935, and finally in the Trade and Navigation Treaty of 1940, was restricted to the Caspian coastal states. The last one provided for the equal treatment of all vessels operating under the flag of the contracting parties both entering and calling at port facility (Article 12). In addition, the parties agreed that no extra fees may be charged to any ship of another contracting party: they can be charged only with fees that are paid by one's own ships.

The bilateral agreements between the USSR and Persia/Iran concluded in 1921 and 1940 became the final legal basis regulating the rights and obligations of coastal

---

[31] Article XIV Treaty of Friendship 1921.

states regarding shipping in the Caspian Sea. After the collapse of the Soviet Union, despite frequent denials of the legally binding nature of these Soviet–Iranian treaties by the newly independent states, their core regulation on the freedom of navigation was also retained and treated as a starting point for negotiations on the future legal status of the Caspian Sea.

9.4 Navigation in the Interstate Negotiations According to the Draft Convention on the Caspian Sea Legal Status

The draft of the future convention on the legal status of the Caspian Sea regulated navigation as follows:

> Merchant ships flying the flag of a contracting state enjoy the freedom of navigation on the entire Caspian Sea. The freedom of commercial navigation on the Caspian Sea is exercised according to the provisions of this Agreement and other agreements of contracting parties, which remain in accordance with this treaty (Article 10(10), Abs. 1).

This provision on the freedom of navigation on the Caspian Sea reflected UNCLOS's provisions regarding the high seas, albeit in a largely modified form. According to UNCLOS, the freedom of navigation[32] constitutes one of the main "maritime freedoms," and it served as a basis for the principle of freedom of the high seas. Thus, within the freedom of each state, whether coastal or landlocked, was the right to sail ships flying its flag on the high seas.[33] The Draft Caspian Status Convention grants unrestricted rights to navigate on the entirety of the Caspian Sea, without considering the existence of any special maritime zones. The discussions over the future delimitation of the Caspian Sea have not yet arrived at whether, and if so to what extent, the coastal states will exercise sovereign rights in future maritime zones in the Caspian Sea. However, regardless of the outcome of this dispute, the legal status of respective zones will not impact the coastal states' freedom of navigation in the entire Caspian basin. The Draft Caspian Status Convention did not follow the traditional distinction included in UNCLOS between the regime of navigation within the internal waters, the territorial seas, and the exclusive economic zone.

The Draft Caspian Status Convention Sea provides for the national treatment of all vessels as follows:

> Each State Party shall guaranty merchant ships of other contracting parties the same treatment as to the national merchant ships. This includes unrestricted calling in at national

---

[32]Freedom and Safety of Navigation out of coastal sea area is regulated by UNCLOS (Part XII, Section 7) and other conventions like Convention on the International Regulations for Preventing Collisions at Sea and other rules set by International Maritime Organisation, which role is to impartially monitor navigational routs.

[33]Article 90 et seq. UNCLOS.

ports in the Caspian Sea for the purpose of loading and unloading of the cargo, the embarkation and disembarkation of passengers, payment of shipping and other port charges, as well as the use of the ordinary for shipping and carrying out of particular services to commercial activities (Article 10(10), Para. 2).

The benefits provided here were limited by the following rule:

> The regime shown in paragraph 2 of this Article shall be applicable to the ports of the Caspian Sea, which are open to the vessels flying the flag of States Parties (Article 10(10), Para. 3).

The above principle of national treatment means a ban on discrimination against national ships. It confirmed that all benefits granted to a merchant ship flying the flag of a contracting state must be granted also to a merchant ship flying flags of all other states parties.

The national treatment clause set by the Soviet–Iranian treaties remains in line with the Draft Caspian Status Convention and the future Convention's principles concerning the equality clause.

The following rules regarding the transit rights on inland waterways for the Caspian states were included into the Draft Caspian Status Convention:

> Conditions and Procedure of the transit from the ocean through the internal waters of Russia for vessels flying the flag of Azerbaijan, Iran, Kazakhstan or Turkmenistan are to be set in an agreement between any of these countries and the transit state (Article 10(10), Abs. 4–5, proposed by Russia, Iran, and Turkmenistan)

or

> The Republic of Azerbaijan, the Republics of Kazakhstan and Turkmenistan as landlocked states have the right to free access to other seas and the ocean. For this purpose they exercise the freedom of transit with all means of transport through the territory of the Islamic Republic of Iran and the Russian Federation (Article 10(10), Abs. 4–5, proposed by Azerbaijan and Kazakhstan)

or

> The Contracting Parties, which are landlocked, have the right of access to other seas and to the ocean. For this purpose they exercise the freedom of transit with all means of transport through the territory of transit countries. Conditions and Procedure of exercising the freedom of transit are to be set in bilateral, sub-regional or regional agreements between any of these countries and the transit state (Article 10(10), Abs. 4–5, proposed by Kazakhstan, supported by Azerbaijan).

There was, however, no consent among the Caspian littoral states regarding the conditions of possible access by Azerbaijan, Kazakhstan, and Turkmenistan to the other seas and the oceans. Depending on whether the negotiating state was a landlocked country itself or not.

In the early 1990s, Kazakhstan, in representing the first option, used to call upon the states bordering the Caspian Sea, according to Part IX of UNCLOS, to cooperate with each other in the exercise of their rights and in the performance of their duties (Article 123). This approach, without referring directly to UNCLOS, was still represented by the Caspian's three landlocked countries. In contrast, Russia, backed by Iran and Turkmenistan, represents the view that the future standards regulating the freedom of transit for the landlocked Caspian countries shall be settled in a special agreement between the landlocked state and the transit state. However, Russia often wavered in its opinion on the matter.

With regard to the right of passage for nonmerchant vessels in the Caspian Sea, the draft of the future Caspian status convention provided the following:

> Warships and other government ships operated for non-commercial purposes enjoy the right of transit through the zones of national jurisdiction of other states Parties. The passage must be continuous and expeditious. However, passage includes stopping and anchoring as long as a tentative agreement exist or are rendered necessary by force majeure or distress or for the purpose of rendering assistance to persons, ships or aircraft in danger or distress (Article 11 (10), proposed by Russia).

Here the proposed rule on the right of innocent passage favored warships and other government ships operated by the contracting states for noncommercial purposes. It followed only partly the provisions of UNCLOS related to the right of innocent passage through the territorial sea by ships of all states (Article 17 et seq.).

Another difference between Russia's proposal and the UNCLOS provisions was that Russia requests peaceful passage within the zones of national jurisdiction, not the territorial sea. Russia views free passage as reserved exclusively to certain contracting states, rather than recognizing this right as belonging to all states, as UNCLOS did. Stopping shall be allowed merely in emergencies for persons, ships, or aircraft.

The Draft Caspian Status Convention provides for states' jurisdiction over their nationals:

> State Parties shall exercise their sovereignty [Russia proposed removal of this notion] and their jurisdiction in the Caspian Sea over their nationals, their ships, installations and structures according to the norms of international law (Article 12(11), Section 1).

Finally, the Draft Caspian Status Convention provided for safety zones in the Caspian Sea:

> Geographical coordinates of the structures and contours of the safety zones shall be communicated to all contracting states (Article 12(11), Section 2).

The above proposal remained in accordance with the standards of international management of safety zones.

9.5 Navigation Regime According to the Convention on the Caspian Sea from 2018

Due to the enormous importance of navigation on the Caspian Sea as a carrier of trade and economic development of the riparian states, during the interstate negotiations after the collapse of the Soviet Union, the riparian states discussed a framework for a special regional regulation on navigation. The regime reflected in the draft of the Caspian Sea Convention discussed among the coastal states since the early

1990s remained far from the general previsions of the law of the sea. No willingness of application of law of the sea standards to the Caspian Sea was justified by the fact that Caspian Sea is not a "sea" and therefore there would not be a necessity to apply international legal standards to the Caspian navigation.

Navigation is to be seen as one of the areas where the norms of international law are to enjoy a definite primacy over the regional regulations.[34] The national legislation of coastal states may not violate internationally accepted standards applicable in specific cases.

9.5.1 Freedom of Navigation in the Caspian Sea

With regard to the regulation of navigation in the Caspian Sea, the indirect applicability of UNCLOS shall be considered.

The Caspian Sea Convention regulates the legal regime of the Caspian Sea as follows:

> Ships flying the flags of the Parties shall enjoy freedom of navigation beyond the outer limits of the Territorial Waters of the Parties. The freedom of navigation shall be exercised in accordance with the provisions of this Convention and other compatible agreements between the Parties without prejudice to the sovereign and exclusive rights of the Parties as determined in this Convention (Article 10, Para. 1).
>
> Each Party shall grant ships flying the flags of other Parties that carry goods, or passengers and baggage, or perform towing or rescue operations the same treatment as to its national ships with regard to free access to its ports in the Caspian Sea, their use for loading and unloading goods, boarding or disembarking passengers, payment of tonnage and other port dues, use of navigation services and performance of regular commercial activities (Article 10, Para. 2).
>
> The regime specified in Para. 2 of this Article shall apply to the ports in the Caspian Sea that are open to ships flying the flags of the Parties (Article 10, Para. 3).

The newly introduced navigation regime in the Caspian Sea reflects some of the UNCLOS standards on navigation, albeit in a largely modified form. The freedom of shipping applies, and is acknowledged, beyond the outer limits of the territorial waters, being under national sovereignty (Article 10, Caspian Sea Convention). This freedom shall be, however, exercised in accordance with the Convention's provision, which imposes significant restrictions on navigation. On one side, Article 10, Para. 2, contains provisions on cargo and harbor access, as well as nondiscrimination treatment on harbors with regard to fees, charges, etc. In other words, it forbids flag discrimination practices, which otherwise would take effect in zones under the sovereignty of the coastal states. The principle of equal treatment is deeply rooted in the navigation tradition of the Caspian Sea states. The agreements concluded by the Soviet Union with Iran established the equal treatment of all

---

[34]See: (Vukas 2004), p. 133.

9.5 Navigation Regime According to the Convention on the Caspian Sea from 2018

vessels flying the flag of the states parties.[35] On the other side, the navigation rights over the Caspian Sea outside of the territorial waters of the coastal states are restricted according to Article 12, Para. 3. It says that

> each Party's, in the exercise of its sovereignty, sovereign rights to the subsoil exploitation and other legitimate economic activities related to the development of resources of the seabed and subsoil, and exclusive rights to harvest aquatic biological resources as well as for the purposes of conserving and managing such resources in its Fishery Zone, may take measures in respect of ships of other Parties, including boarding, inspection, hot pursuit, detention, arrest and judicial proceedings, as may be necessary to ensure compliance with its laws and regulations.

According to this article, in both common maritime space and fishery zones, navigation rights may be clearly limited by the coastal states. The reason for justifying a recourse to such strong measures of intervention in the navigation of the Caspian Sea may be explained by a state's sovereign rights over the exploitation of the resources in the sectors extending over the bottom of the Caspian Sea. The application of the measures stipulated in this paragraph needs to be justified. In the event that such restrictive measures over ships are left unjustified, there should be compensation for any losses and damages that occur. Such an approach to the freedom of navigation is not widely known in the international law of the sea. UNCLOS allows the coastal state in the EEZ, in the exercise of its sovereign rights to explore, exploit, conserve, and manage the living resources in the exclusive economic zone, to take measures such as boarding, inspection, arrest, and initiation of judicial proceedings (Article 73, Para. 1, UNCLOS). On the other side, UNCLOS provides for the freedom of navigation on the high sea (Article 87), where such restrictive measures as boarding, hot pursuit, etc. are outlined merely under narrowly specified circumstances.

Following the standards set up in UNCLOS, the Caspian Sea Convention does not expand the freedom of navigation to the territorial waters (Article 10, Para. 1). Areas with the most shipping routes are located along coastal areas, where riparian states exercise their sovereignty. To allow for international navigation and trade, the law of the sea limits the authority of the coastal states in the territorial zones (Article 18, UNCLOS).[36] Also, Article 11, Para. 1, of the Caspian Sea Convention allows "vessels, flying the flags of the other coastal states, for navigation through Territorial Waters with a view to traversing those waters without entering Internal Waters or calling at a roadstead or port facility outside Internal Waters or proceeds to or from Internal Waters."

---

[35] See: Article XV of 1945 Treaty; Article XII of 1940 Treaty.
[36] Wolfrum (1990).

9.5.2 Innocent Passage in the Caspian Sea

Similarly to UNCLOS, Article 21 of the Caspian Sea Convention defines laws and regulations of the coastal states relating to innocent passage. Accordingly, the Caspian Sea Convention states:

> A Party shall duly publish all laws and regulations related to the passage through its Territorial Waters (Article 11, Para. 9).

The right of innocent passage was reserved exclusively for the contracting parties and not extended to other countries, which would be necessary in case of application of UNCLOS standards to the regulations of the Territorial Seas. The right of innocent passage in the Caspian Sea Convention is defined as follows:

> 1. Ships flying the flags of the Parties may navigate through Territorial Waters with a view to:
>
> a) Traversing those waters without entering Internal Waters or calling at a roadstead or port facility outside Internal Waters;
>
> or
>
> b) Proceeding to or from Internal Waters or calling at such roadstead or port facility (Article 11, Para. 1).

Further on, according to the Caspian Sea Convention,

> passage through Territorial Waters must not be prejudicial to the peace, good order or security of the coastal State. Passage through Territorial Waters should be continuous and expeditious (Article 11, Para. 3).

It exactly follows the regulation of UNCLOS (Article 18, Para. 2), except that UNCLOS requires the passage to take place in conformity also with rules of international law, whereas the Caspian Sea Convention refers only to its own provision as a standard for the assessment of innocent passage. Further on, the Caspian Sea Convention envisages the following:

> a) A Party may adopt laws and regulations, in conformity with provisions of this Convention and other norms of international law, relating to passage through Territorial Waters, including in respect of all or any of the following:
> b) Safety of navigation and regulation of maritime traffic;
> c) Protection of navigational aids and facilities, as well as other facilities or installations;
> d) Protection of cables and pipelines;
> e) Conservation of the biological resources of the sea;
> f) Prevention of infringement of fishery laws and regulations of the coastal State;
> g) Preservation of the environment of the coastal State and prevention, reduction and control of pollution thereof;
> h) Marine scientific research and hydrographic surveys;
> i) Prevention of infringement of customs, fiscal, immigration or sanitary laws and regulations of the coastal State;
> j) Ensuring national security (Article 11, Para. 8).

9.5 Navigation Regime According to the Convention on the Caspian Sea from 2018

The Caspian Sea Convention provides for sea-lanes and traffic separation schemes in the territorial sea but in a more limited scope than that of UNCLOS (Article 22):

> Each Party may, where necessary and with due regard to the safety of navigation, require ships flying the flags of other Parties passing through its Territorial Waters to use such sea lanes and traffic separation schemes as it may designate or prescribe for the regulation of the passage of ships through its Territorial Waters (Article 11, Para. 11).

In terms of the duties of the coastal state, the Caspian Sea Convention fully follows the provisions of Article 24, Para. 1, UNCLOS:

> A Party shall not hamper the passage of ships flying the flags of other Parties through its Territorial Waters except when acting under this Convention or laws and regulations adopted in conformity therewith. In particular, a Party shall not:
>
> a) Impose requirements on ships flying the flags of other Parties which have the practical effect of denying or impairing unjustifiably the right of passage through its Territorial Waters;
>
> or
>
> b) Discriminate in form or in fact against ships flying the flags of other Parties or ships carrying cargoes to, from or on behalf of any State.
>
> A Party shall give appropriate publicity to any danger to navigation within its Territorial Waters of which it has knowledge (Article 11, Para. 16).

In terms of the right of protection of the coastal state, the Caspian Sea Convention envisages the following regulations, which remains in conformity with Article 25, Paras. 1, 2, 3, of UNCLOS:

> In its Territorial Waters, a Party may take necessary steps to prevent a passage through its Territorial Waters violating conditions set forth in this Article (Article 11, Para. 7).
>
> In the case of ships proceeding to Internal Waters or a call at port facilities outside Internal Waters, the coastal State shall also have the right to take necessary steps to prevent any breach of the conditions to which admission of those ships to Internal Waters or such a call is subject (Article 11, Para. 12).
>
> A Party may, without discrimination in form or in fact against ships flying the flags of other Parties, temporarily suspend in specified areas of its Territorial Waters the passage of those ships if such suspension is essential for the protection of its security (Article 11, Para. 13).
>
> Ships flying the flags of the Parties, while passing through Territorial Waters, shall observe all laws and regulations of the coastal State related to such passage (Article 11, Para. 10).
>
> Each Party may, where necessary and with due regard to the safety of navigation, require ships flying the flags of other Parties passing through its Territorial Waters to use such sea lanes and traffic separation schemes as it may designate or prescribe for the regulation of the passage of ships through its Territorial Waters (Article 11, Para. 11).

The terms and conditions of innocent passage within the territorial waters of the Caspian Sea are comparable to international law standards with respect to the innocent passage of warships. As indicated already before, the right of the innocent passage was reserved exclusively to the Caspian states and by that differs from the standard included in the regulations of UNCLOS for the Territorial Seas.

Rules applicable to warships and other governmental ships operating for noncommercial purposes differ from UNCLOS (Section C) regulations and will be discussed separately in the following subchapter.

The terms and conditions of innocent passage within the territorial waters of the Caspian Sea are comparable to international law standards with respect to the innocent passage of warships.

9.5.3 Nonmerchant Vessels

With regard to the passage of nonmerchant vessels in the Caspian Sea, the Caspian Sea Convention provides for the following regime, which only partly follows the international law of the sea, as included in Section C of UNCLOS. The definition of warships in the Caspian Sea Convention is the same as the UNCLOS definition:

> Warship – a ship belonging to the armed forces of a Party and bearing external marks distinguishing such ships of its nationality, under the command of an officer duly commissioned by the government of the Party and whose name appears in the appropriate service list or its equivalent, and manned by a crew which is under regular armed forces discipline (Article 1).

However, the Caspian Sea Convention does not fully reflect the provisions of UNCLOS, which in Article 32 speaks of immunities of warships and other government ships operated for noncommercial purposes; neither does it reflect Article 18 of UNCLOS, which provides for the meaning of innocent passage. The Caspian Sea Convention envisages the following:

> Terms and procedures for the passage of warships, submarines and other underwater vehicles through Territorial Waters shall be determined on the basis of agreements between the flag State and the coastal State or, in the absence of such agreements, on the basis of the coastal State legislation. If the entry of a warship to the Territorial Waters is required due to a force majeure or distress, or to render assistance to persons, ships or aircraft in distress, the captain of the warship approaching the Territorial Waters shall notify the coastal State accordingly, with such entry to be performed along the route to be determined by the captain of the warship and agreed with the coastal State. As soon as such circumstances cease to exist, the warship shall immediately leave the Territorial Waters. Terms and procedures for the entry of warships to the Internal Waters due to a force majeure or distress, or to render assistance to persons, ships or aircraft in distress shall be determined on the basis of agreements between the flag State and the coastal State or, in the absence of such agreements, on the basis of the coastal State legislation (Article 11, Para. 2).

Further on, the Caspian Sea Convention states the following:

> If a warship or a government ship operated for non-commercial purposes does not comply with laws and regulations of the coastal State concerning passage through Territorial Waters and disregards any request for compliance therewith which is made to it, the Party concerned may require it to leave its Territorial Waters immediately (Article 11, Para. 14).

It reflects the stipulation included in UNCLOS Article 30, which regulates noncompliance by warships with the laws and regulations of the coastal state.

Finally, the Caspian Sea Convention provides:

> The flag Party shall bear international responsibility for any loss or damage to another Party resulting from the non-compliance by a warship or other government ship operated for non-commercial purposes with laws and regulations of the coastal State concerning passage through its Territorial Waters, entering such waters and anchoring therein or with provisions of this Convention or other norms of international law (Article 11, Para. 15).

This provision is in agreement with UNCLOS, Article 31, which notes the responsibility of the flag state for the damage caused by a warship or other government ship operated for noncommercial purposes.

9.5.4 Right to Access the Ocean

An important issue pertaining to navigation and trade to and from the Caspian Sea is the possibility of transit for the landlocked Caspian Sea countries to the ocean. Volga, as mentioned before, is the only conceivable waterway that could be used as a transit route from the Caspian Sea. Since the Volga is an internal watercourse of Russia, Russia shall be recognized as a "transit state" through whose territory the transit of persons and goods would need to happen (Article 124, UNCLOS). The new Caspian Sea Convention confirms that parties shall have the right of free access from the Caspian Sea to other seas and the Ocean, and back. To that end, the parties shall enjoy the freedom of transit for all their means of transport through the territories of transit parties (Article 10, Para. 4, Caspian Sea Convention). However, the terms and procedures for such access shall be determined by bilateral agreements between the parties concerned and the transit parties. According to the Caspian Sea Convention, the transit Parties shall be entitled to take all necessary means to ensure their rights and the possibility to exercise full sovereignty. Such a requirement for earlier approval by the coastal state, or at least the duty of announcing the passage, has never been accepted in the Law of the Sea.[37] Volga belongs, however, entirely to the Internal Waters of Russia. Russia, therefore, should be recognized as a "transit state" (Article 124, UNCLOS). The Caspian Sea Convention, however, requires Caspian Sea states to conclude separate relevant agreements between the transit party and the interested landlocked countries, which are, respectively, Russia, Kazakhstan, Azerbaijan, and Turkmenistan.

---

[37] Article 21 UN Document A/AC 138/SC II/L.18. See also in (Vukas 2004), p. 141.

9.6 Conclusion

Shipping is traditionally one of the most important regimes of the use of the Caspian Sea. It was once used for transporting all kind of goods, but nowadays it is used especially for shipping natural resources. The regulation of shipping was a subject of the earliest establishment of interstate law. The initial agreements concluded by the Soviet Union and Iran provided for the freedom of shipping in the entire Caspian Sea for all ships of the coastal states exclusively.

The newly adopted Caspian Sea Convention in 2018 has recognized coastal states' shipping rights with limitation, both in the internal waters and in the territorial waters, by introducing the concept of innocent passage of the Caspian riparian states. Further, the freedom of shipping was recognized beyond the outer limits of the territorial waters, which shall be exercised in accordance with this Convention. It imposes, however, significant restrictions on navigation-related rights due to the coastal states' right to use the natural resources of the seabed and subsoil in the national sectors of the Caspian Sea. It envisages the right of the coastal states to take measures in respect of ships of other parties, including boarding, inspection, hot pursuit, detention, arrest, and initiation of judicial proceeding, in both the fishery zone and common maritime space. The Caspian Sea Convention has recognized the possibility of transit of the landlocked Caspian Sea countries to the ocean through the Volga. It states, however, that the terms and procedures for such access shall be determined by bilateral agreements between the parties concerned and the transit party.

References

Arsenov W (2003) Internationalee Transportkorridor Nord-Süd. Iran Heute 2
Bruce F (1993) International shipping: an introduction to the policies, politics, and institutions of the maritime world, 33rd edn. Lloyd's of London Press, London
Farthing B, Brownrigg M (1993) Farthing on international shipping, 3rd edn. Lloyd's List Practical Guides
Fox JR (1992) Dictionary of international and comparative law. Oceana publications, Dobbs Ferry
Goss R (1985) Economics and the international regime for shipping. In: Buttler WE (ed) The law of the sea and international shipping. Ocean Publications, New York
Ipsen K (2004) Völkerrecht, 5th edn. CH Beck, München
Kasoulides G (1992) Jurisdiction of the coastal state and regulation of shipping. Revue Hellenique de Droit International 45:300
Lagoni R (1988) Der Hamburger Hafen und die Internationale Handelschiffahrt im Völkerrecht. 26 Archiv des Völkerreichts 355 ed. Ports in International Law, Oxford
Łukaszuk L (1997) Mie̦dzynarodowe Prawo Morza (International law on the sea). Scholar, Warsaw

Tuschin J (1978) Russkoje Moreplavanie na Kaspijskom, Azovskom i Chornom More (Russian sailing in Caspian, Azov & Black Seas). Nauka, Moscow

Uprety K (1995) Right of access to the sea of land-locked states: retrospect and prospect for development. J Int Leg Stud 21 (winter)

Vukas B (2004) The law of the sea: the selected writings. Publications on Ocean Development, 45 edn. Martinus Nijhoff

Wolfrum R (1990) Die Umsetzung des Seerechtsübereinkommens in Nationales Recht. United Nations (б.м.)

Chapter 10
Protection of the Marine Environment of the Caspian Sea

10.1 Introduction

Economic activities related to the Caspian Sea threaten its fragile environment. Overexploitation of resources, destruction of habitats, and pollution from different sources, including from shipping and the development of seabed resources, as well as pipeline transportation of oil and gas, make the environmental picture of the Caspian Sea very problematic. The purpose of this chapter is to examine the significance of the existing regulations on the protection of the fragile Caspian maritime environment. The adequacy of the existing rules will be judged by their ability to protect the marine environment of the Caspian Sea. This analysis presents a rather practical approach to examining existing legal acts in the Caspian Sea and is based on an analysis of and a comparison with the related international treaties and agreements.

Reflecting purely on the growing concerns over the environmental conditions of the Caspian Sea, the coastal states have tried to take appropriate measures to prevent the further destruction of its ecosystem and to preserve it in its best possible condition. The Caspian Sea Convention, adopted in 2018, stresses in its Preamble states' awareness of their responsibility toward the present and future generations for the preservation of the Caspian Sea and the sustainable development of the region. The Convention further confirms the states' conviction that the Convention will facilitate the protection and conservation of the Caspian Sea environment. In its text, the Convention mentions the protection of the Caspian Sea environment, its conservation and restoration, and the rational use of its biological resources as some of the Convention's main principles (Article 3, Para. 14), which shall guide state activities in the Caspian Sea. The Caspian Sea Convention also provides for the liability of the polluting party for damage caused to the environment of the Caspian Sea (Article 3, Para. 13). The Convention defines pollution as follows:

> the introduction by man, directly or indirectly, of substances, organisms or energy into the ecological system of the Caspian Sea, including from land-based sources, which results or is

likely to result in deleterious effects, such as harm to biological resources and marine life, hazards to human health, hindrance to marine activities, including harvesting of aquatic biological resources and other legitimate uses of the sea, impairment of quality for use of sea water and reduction of amenities (Article 1).

The Caspian Sea Convention also specifically prohibits pollution from pipelines (Article 14, Para. 2) and vessels in the territorial sea (Article 11, Para. 6). In its Article 15, the Caspian Sea Convention compiles most of its environment-related regulations and states:

> Parties shall undertake to protect and preserve the ecological system of the Caspian Sea and all elements thereof (Para. 1); the Parties shall take, jointly or individually, all necessary measures and cooperate in order to preserve the biological diversity, to protect, restore and manage in a sustainable and rational manner the biological resources of the Caspian Sea, and to prevent, reduce and control pollution of the Caspian Sea from any source (Para. 2); any activity damaging the biological diversity of the Caspian Sea shall be prohibited (Para. 3); the Parties shall be liable under the norms of international law for any damage caused to the ecological system of the Caspian Sea (Para. 4).

The Caspian Sea Convention reinforces the more detailed provisions relating to the Caspian environment included in the Tehran Convention, which is the main source of regulation on the environmental protection of the Caspian Sea and which will be discussed throughout this chapter. Reflecting on the structure of the most important act providing for the protection of the Caspian environment—the Tehran Convention—this chapter has been divided into parts, presenting the main obligations of coastal states toward the Caspian Sea environment. It begins with the elaboration of the general environmental principles applicable to the Caspian Sea. The next part reflects on the responsibility of states for the prevention, reduction, and control of pollution from various sources, such as land-based, seabed activities, dumping; shipping; introduction of alien species; and other human activities and environmental emergencies that might cause pollution to the Caspian Sea environment. The subsequent part discusses the obligations of states parties to protect, preserve, and restore the marine environment of the Caspian Sea, including its biodiversity and the management of the coastal zone, as well as the effect of the fluctuation of Caspian Sea's level. Further, institutional arrangements and a number of special procedural instruments of maritime protection are discussed, such as exchange of information, environmental impact assessment (EIA), state cooperation, monitoring, research and science, consultations, and access to information for the public. Finally, the issue of implementation of the Convention and compliance with its provisions, via liability and compensation provisions as well as mechanisms of dispute settlement, are discussed. The elaboration made here on the states' obligations toward the Caspian Sea environment is based mainly on the provisions of the Tehran Convention but is expanded to include the contents of its ancillary protocols. The provisions of the Tehran Convention regulate a complexity of environmental phenomena in the Caspian region that are caused by a variety of factors influencing the ecosystem of the region. Coping with them requires holistic cooperation and understanding among all coastal states, which in the Caspian case are often undermined by the lack of political will to limit one's own sovereign powers and

10.1 Introduction

to offer sufficient financial means to cover environmental needs. As the awaited environmental solution for the Caspian Sea seems most likely to be reached in a gradual process, the coastal states have decided to reach an objective of protecting all spheres of the marine environment of the Caspian Sea by a number of instruments to be concluded one by one. A similar trend can be observed in the present international practice of responding to global environmental challenges.[1] The Tehran Convention alone would hardly develop a practical effect, except that it obliges its states parties to undertake certain further actions. With the adoption of the Tehran Convention, the states parties set specific environmental goals but avoided taking on explicit commitments. A legally binding effect can only be achieved through the adoption of implementing protocols, something that takes place gradually. Up to date, three additional protocols to the Tehran Convention—the Aktau Protocol (2011), the Moscow Protocol (2012), and the Biodiversity Protocol (2014)—have been signed by the Caspian countries. However, they have not yet entered into force. Additional work is underway to prepare the Protocol on Environment Impact Assessment in a Transboundary Context (EIA Protocol). Apart from the Tehran Convention, two more environmental documents that add to the general understanding of the framework of protection of the Caspian Sea environment are under the Caspian states' consideration. The first one was prepared under the auspices of the Commission on Aquatic Bioresources of the Caspian Sea as a Draft Agreement on Conservation of Aquatic Bioresources of the Caspian Sea and Their Management. It includes the main principles and criteria of management of the aquatic bioresources stock of the Caspian Sea, which have been discussed among the states since 2003. If this document was finally adopted by all Caspian states and entered into force, there would be no need for an extensive regulation of this issue within the Protocol on Conservation of Biological Diversity ancillary to the Tehran Convention, which has, however, been adopted in May 2014. Second, the Agreement on the cooperation among Caspian states in the area of hydrometeorology of the Caspian Sea has been proposed by the Coordinating Committee on Hydrometeorology and Pollution Monitoring of the Caspian Sea (further referred to as CASPCOM) during its 17th session in October 2012[2] but has not been agreed yet. (See further details in Sect. 10.7.1.) Both documents are still under the states' consideration and have yet to reach a final legal form. Meanwhile, the Agreement on cooperation in the field of prevention and liquidation of emergency situations was already adopted on September 29, 2014, and entered into force on September 19, 2017 (for further details, see Sect. 4.5.3).

---

[1] See: (Land 2000), p. 108.
[2] Sea (2005).

10.2 Protocols for the Enforcement of the Tehran Convention

The creation of a tailor-made, mutually beneficial regime for all the sensitive issues of the marine environment of the Caspian Sea was a difficult undertaking as the states have very heterogeneous interests resulting from their economic needs. To facilitate the negotiation process on the Tehran Convention for the Protection of the Marine Environment of the Caspian Sea, the coastal states decided to tackle the existing problems gradually and design a framework agreement to be fulfilled through additional detailed protocols. As the adoption of auxiliary protocols to the Tehran Convention is both a legal obligation provided in the Convention itself and a necessary condition for the successful implementation of the Tehran Convention, it seems important to elaborate on this requirement:

> Any Contracting Party may propose Protocols to this Convention. Such Protocols shall be adopted by unanimous decision of the Parties at a meeting of the Conference of the Parties. Protocols shall enter into force after their ratification or approval by all the Contracting Parties in accordance with their constitutional procedures, unless the Protocol does not envisage a different procedure for adoption. Protocols shall form an integral part of this Convention (Article 24(1)).

> The annexes to this Convention or to any Protocol shall form an integral part of the Convention or of such Protocol, as the case may be, and, unless expressly provided otherwise, a reference to this Convention or its Protocols constitutes at the same time a reference to any annexes thereto. Such annexes shall be restricted to procedural, scientific, technical and administrative matters (Article 25(1)).

The working version of the Tehran Convention and the drafts of all the additional protocols were prepared on behalf of the governments of the Caspian littoral states by the United Nations Environmental Programme (UNEP). UNEP has a long record of supporting numerous processes of negotiating multilateral environmental agreements on regional level.[3] The Caspian Environmental Programme (CEP) took over the organizational tasks and mediation between UNEP and the negotiating Caspian governments. The approach of complementing an international treaty by a series of additional instruments aims at both facilitating difficult negotiations and reducing the number of any future amendments to the text of the Tehran Convention. In the case of any amendment to the Tehran Convention, the procedures are rigorous. The Tehran Convention prohibits reservations (Article 32). It means that any unilateral action by coastal states, which purport to exclude or to modify thelegal effect of certain provisions of the treaty is not allowed.[4]

---

[3]UNEP Regional Seas Programme; Washington Global Programme of Action for Protecting the Marine Environment from Land-Based Activities 1995; Rotterdam Convention 1998.

[4]Definition of reservations according to Article 2(1)d of the Vienna Convention on the Law of Treaties.

The Tehran Convention appoints no fixed time frame to the Conference of the Parties for the adoption of implementing protocols. It determines neither the precise objectives to be achieved with these protocols nor the order of their adoption. However, it seems that the protocols must be accurate enough so as not to miss the target of complementing the provisions of the Tehran Convention. Initially, the Protocol Concerning Regional Preparedness, Response and Cooperation in Combating Oil Pollution Incidents (the "Aktau Protocol") was adopted on August 12, 2011, and entered into force on July 25, 2016. Next, the Protocol for the Protection of the Caspian Sea Against Pollution from Land-Based Sources and Activities (the "Moscow Protocol") was adopted on December 12, 2012, but has not been ratified yet. The Protocol for the Conservation of Biological Diversity (the "Ashgabat Protocol") was adopted in Ashgabat, Turkmenistan, only on May 30, 2014, and its entry into force depends on its ratification by the remaining Caspian Sea states. The Protocols on Environmental Impact Assessment (EIA) in a Transboundary Context was adopted on July 20, 2018, though it is not yet in force.

10.3 Environmental Principles Applicable to the Caspian Sea

The provisions of the Tehran Convention are rooted in the well-established principles of international environmental law. The Tehran Convention explicitly refers to the fundamental principles of precaution, cooperation, sustainability, responsibility, liability, etc. Some of them enjoy a binding legal force, and some others need to be complemented by other norms.

10.3.1 Principle of Sustainable Development

According to the Tehran Convention (Article 2), the objective of the treaty is to use the resources in a rational way. This goal reflects the legal principle of sustainable development, which is to be regarded as one of the basic rules of modern international environmental law. It has also been named in the Moscow Protocol (Preamble) to the Tehran Convention. The EIA Protocol mentions this principle, recognizing that the application of an environmental impact assessment at an early stage in the decision-making process promotes the implementation of the principle of sustainable development (Preamble). Also, the Ashgabat Protocol echoes this principle in Article 1 in the context of sustainable use of the components of biological diversity. In the Caspian Sea Convention, the parties highlighted their awareness of their responsibility for the sustainable development of the region (Preamble), as well as acknowledges their commitment to jointly or individually cooperate for the sustainability of the biological resources of the Caspian Sea. The emphasis on the principle of sustainable development by the Caspian littoral states confirms their commitment to the right of human community to live in uninjured environmental conditions, reflecting the eco-friendly awareness of the countries of the Caspian region, which will be beneficial for the further development of environmental cooperation in the region. The recognition of interaction between the environment and development derives from the Stockholm Declaration[5] and, in particular, from the Final Declaration of the United Nations Conference on Environment and Development, signed in 1992 in Rio de Janeiro.[6] Since 2015, sustainable development has become a key aspect for the new 2030 agenda for sustainable development adopted by United Nations member states and its 17 sustainable development goals (SDGs).[7] It is arguable whether the principle of sustainable development has already become part of customary international law. However, certainly the number of international agreements recognizing this principle is increasing. As not one generally accepted definition of sustainability exists, some treaties refer to it by using other notions like "rational,"[8] "proper,"[9] or "wise."[10] A sustainable use of natural resources aims at achieving the environmentally and socially acceptable forms of the use and exploitation of the naturalresources for the purpose of economic development. The application of this principle in environmental legal praxis that requires in-depth knowledge in the field of natural sciences to create an appropriate legal regime for the exploitation of natural resources.[11]

Sustainable development of natural resources means, to simplify it a bit, the reduction of resource consumption to a level that does not exceed the regenerative capacity of this resource's potential. Here, the three dimensions of ecology,

---

[5]Text in: ILM 11 (1972), 1416.

[6]Text in: ILM 31 T.

[7]Sustainable Development Website <https://sustainabledevelopment.un.org> accessed July 18, 2020.

[8]Stockholm Declaration 1972, Principle 13 and 14; Antarctic Marine living Resources Convention 1980, Article II(1) and (2); Jeddah Convention 1982, Article 1(1) define "conservation" objectives as including "rational use," regarding migratory birds: Western Hemisphere Convention 1940, Article VII]; regarding fisheries: Danube Fishing Convention 1958, Preamble and VIII; 1959 North-East Fisheries Convention, Preamble and Article V(1)(b); Black Sea Fishing Convention 1959, Preamble and Article 1 and 7; South Atlantic Fisheries Convention 1969, Preamble; Baltic Fishing Convention 1973, Article I and X(h); Northwest Atlantic Fisheries Convention 1978, Article II(1), regarding salmon: North Atlantic Salmon Convention 1982, Preamble, regarding all natural resources: 1968 African Conservation Convention, Article II, Amazonian Treaty 1978, Articles I and VII, regarding seals: Antarctic Seals Convention 1972, Article 3(1); Convention on Conservation of North Pacific Fur Seals 1976, Article II(2)(g), regarding hydro resources Amazonian Treaty 1978, Article V.

[9]Regarding fisheries: General fisheries Council for Mediterranean 1949, Preamble and Article IV (a); regarding forests: American Forest Institute 1959, Article III(1)(a).

[10]Regarding flora and fauna 1968 African conservation Convention, Article VII (1); Stockholm Declaration 1972, Principle 4, South Pacific Nature Conservation 1976, Article V(1); regarding wetlands: Ramsar Wetlands Convention 1971, Articles 2(6) and 6(2)(d); regarding natural resources generally: Bonn Convention 1979, Preamble.

[11]See: Robinson, S. 2 et seq.

economy, and social affairs in the usage of resources must be linked. An important aspect of sustainable development of natural resources is to ensure the respect of needs of the present generation in using natural resources, without compromising the ability of future generations to use such resources.

Under the umbrella of the sustainability principle, various principles have also emerged and are commonly used in international environmental practice, which should guide also the Caspian contracting parties. The Tehran Convention explicitly requires the use of the principles of intergenerational equity, "precautionary," "the polluter pays," and public participation to successfully achieve the objectives of the Tehran Convention and to implement its provisions (Article 5).

10.3.2 *"Future Generations" Principle*

As the Preamble of the Tehran Convention puts it, the contracting parties resolved firmly "to preserve living resources of the Caspian Sea for present and future generations." The Moscow Protocol expresses the states' desire to meet their needs through the protection and conservation of the Caspian environment. The reference to the meeting of the needs and aspirations of the present and future generations is included in Article 1 of the Ashgabat Protocol. Also, in the Caspian Sea Convention, parties expressed their awareness of their responsibility toward the present and future generations for the preservation of the Caspian Sea. The idea of safeguarding natural resources in the interest of future generations and not leaving them resource-scarce and with pollution problems, as an aspect of the sustainable development concept, is reflected in a great number of environmental treaties concerning such issues.[12] Also in international declarations, there are references to the interest of the future generations.[13] The practical legal consequence of provisions concerning this matter is not clear. In general, this principle shall support the possibility of individuals to use their rights and fulfil obligations following from environmental treaties.[14]

In terms of natural resources, it is important to emphasize that natural resources may be developed and managed only to the extent that their long-term usability and availability are ensured also for future generations. The exhaustion of natural resources by the present generation will seriously impact the existence of the future generations.[15] Such a situation would significantly worth the overall life expectation

---

[12]South Pacific Nature Convention 1976, Preamble; 1992 Helsinki Convention, Article 2(5)(c); 1985 ASEAN Convention, Preamble; Kuwait Convention 1978, Preamble; 1983 Cartagena de Indias Protocol, Preamble; Jeddah Convention 1982, Article 1 (1); Biodiversity Convention 1992, Preamble; Climate Change Convention 1992, Article 3(1); Nairobi Convention 1985, Preamble; CITES 1973, Preamble; ENMOD Convention 1977, Preamble; Bonn Convention 1979, Preamble.

[13]UN General Assembly Resolution 35/8 of 1980 Historical Responsibility of States for the Preservation of Nature for Present and Future Generations; Rio Declaration 1992, Principle 3.

[14]United Nations General Assembly Resolution 2749 (XXV) of 17 December 1970.

[15]See: (Thompson 2004).

for future generation and therefore shall be condemned from an ethical and also legal perspective.[16] A possible elimination of the generation conflict can be done through an economic method, the so-called "exponential discounting," which converts a future benefit into an equivalent that corresponds to today's values.[17]

10.3.3 The Precautionary Principle

The precautionary principle is based on the *prevention principle* in German law.[18] The most important feature of this principle is that a positive action to protect the environment may be required before a scientific proof of harm has been provided. However, to this day, there is still no unity in the understanding of the meaning of this term among states. On the one hand, it can be defined as showing caution of approach toward activities that could have an adverse impact on the environment. On the other hand, there could be strict regulation and prohibition of activities and substances potentially harmful to the environment even without any convincing evidence of their likely harmful effect. The precautionary principle is applicable to risk prevention in situations that can be defined as "not-on-risk-yet."[19] The consideration and preparedness for a possible future danger correspond to a long-term environmental perspective that focuses not only on immediate threats. The strengthening of the precautionary approach is reflected in the so-called "cradle to grave" principle, known in American environmental law. It provides for the control of certain environmental problems throughout the process of production, use, and disposal. This approach was applied for the first time in the "Beveridge Report."[20] There is no doubt that the precautionary principle reflects a principle of customary law, and its application will depend on the circumstances of each case. Since the mid-1980s, one can observe a growing support for precautionary actions in binding agreements[21] in the case of a threat of a "serious or irreversible" damage to the environment.[22] The provisions of the Tehran Convention and the Moscow Protocol concerning the precautionary principle are a reiteration of Principle 15 of the Rio Declaration, which is widely considered as the first and full restatement of the precautionary principle. Both of those documents copied almost verbatim the same

---

[16]See: (Davidson 2003).
[17]See: (Farber 2003), S. 1.
[18]Twelfth Report, Royal Commission on Environmental Pollution (1988), p. 57.
[19]See: (Fleury 1995), S. 45.
[20]See: (O'connell and Oldfather 1993).
[21]Vienna Convention 1985, Preamble; Montreal Protocol 1987, Preamble; The Ministerial Declaration of the Second North Sea Conference 1987; The Third North Sea Conference 1990.
[22]Bamako Convention 1991, Article 4(3)(f); Helsinki Convention 1992, Article 2(5)(a); Biodiversity Convention 1992, Preamble; Climate Change Convention 1992, Article 3(3); OSPAR Convention 1992, Article 2(2)(a); 1992 Baltic sea Convention Article 3(2); Rio Declaration 1992, Principle 15.

provision of the Rio Declaration, saying that "where there is a threat of serious or irreversible damage" to the Caspian Sea environment, "lack of full scientific certainty should not be used as a reason for postponing cost-effective measures to prevent" such damage. Also, the Ashgabat Protocol in its Article 2 refers to the precautionary principle of preventing the decline and degradation of and damage to species' habitats and ecological systems.

10.3.4 "The Polluter Pays" Principle

The international environmental law principle known as "polluter pays" principle is reflected in the Tehran Convention:

> In their actions to achieve the objective of this Convention and to implement its provisions, the Contracting Parties shall be guided by, inter alia "the polluter pays" principle, by virtue of which the polluter bears the costs of pollution including its prevention, control and reduction (Article 5(b)).

It is doubtful whether the "polluter pays" principle may be considered as part of internationally binding customary law. It raises no doubts only in relation to state members of the Organisation for Economic Co-operation and Development (OECD)[23] and the EU.[24] The "polluter pays" principle takes its origins from Principle 16 of the Rio Declaration of 1992. It means the polluter should, in principle, bear the cost of pollution. This principle means the cost of removal of the damage caused is attributed to the polluter, and the polluter bears the responsibility of protecting the environment and avoiding causing damage to it. So far, however, the burden of proof has not been clearly defined in international environmental law, nor have the limits of the applicability of the "polluter pays" principle been fixed, although this principle is incorporated into numerous international agreements[25] and nonbinding declarations.[26]

---

[23] OECD Council Recommendation on Guiding Principles Concerning the International Economic Aspects of Environmental Policies C(72)128(1972), 14 I.L.M. (1975), 236; 1989 OECD Council Recommendation on the Application of the Polluter-Pays Principle to Accidental Pollution, C (89) 88(Final)(1989), 28 I.L.M. 1320.

[24] 1973 Programme of Action on the Environment, OJ C 112, 20.12.1973, p. 1; Council Recommendation 75/436/EURATOM, ECSC, EEC of 3 March 1975, Annex, Para. 2, OJ l 169, 29.6.1987, p. 1.

[25] 1985 ASEAN Convention, Article 10(d); 1991 Alpine Convention, Article 2(1); 1992 Helsinki Convention, Article 2(5)(b), OSPAR Convention 1992, Article 2(2)(b); 1992 Baltic Sea Convention, Article 3(4); 1990 Oil Pollution Preparedness Convention, Preamble; Industrial Accidents Convention 1992, Preamble.

[26] Recommendation on the implementation of the Polluter-Pays Principle, C(74)223 (1974); Recommendation on the Application of the polluter-pays principle to Accidental Pollution, C (88)89 (Final)(1989) 28 ILM 1320.

The recognition of this principle and of the obligations that result from it in the Tehran Convention, and in the Moscow Protocol, is especially remarkable. Also, the Caspian Sea Convention states that the parties shall carry out their activities in the Caspian Sea in accordance with the principle of liability of the polluting party for damage caused to the ecological system of the Caspian Sea (Article 3, Para. 13). However, the contracting parties did not positively lose an old subject of contention, whether the polluter, besides paying for the prevention, control, and reduction costs, should also pay for the decontamination, cleanup and restoration of the environment.

On the one hand, the Rio Declaration promotes compliance with the "polluter pays" principle at the international level (Principle 16), but on the other hand it provides for the concept of "common but differentiated responsibility," which acknowledges that taking into consideration the different level of contribution to global environmental degradation states' responsibility for the environmental damage defers (Principle 7). Although the Tehran Convention does not directly refer to the common but differentiated responsibility principle, it leaves open the catalog of environmental principles. The implementation of the "polluter pays" principle requires development of respective national legal framework and cannot be replaced merrily to the reference to the existence of these principles on the level of an international law sources.

10.4 Prevention, Reduction, and Control of Pollution in the Caspian Sea

The sources of maritime pollution are, in particular, pollutants from land, ships, and the air, as well as from oil tankers, activities on the seabed for the exploitation of natural resources, and the disposal of waste of all kinds at sea. One of the main objectives of the Tehran Convention is the prevention, reduction and control of pollution, which constitutes a general obligation of the Caspian Sea littoral states. In the Moscow Protocol, this obligation was extended by the requirement to eliminate pollution to the maximum extent possible by developing new methods and techniques of pollution, prevention, reduction and elimination, as well as cooperating with international organizations on this matter (Article 16). Also, in the EIA Protocol, the parties agreed to carry out environmental impact assessment in order to prevent, reduce, and control the pollution of the maritime environment of the Caspian Sea (Article 2), which was in detail discussed in Sect. 10.7.3. The only protocol in force today, the Aktau Protocol, contains provisions regarding regional response and cooperation in combating oil pollution resulting from operational measures on ships or offshore units, as well as pollution originating from land-based sources (Article 3). The Caspian Sea Convention obliges states parties to carry out all activities in accordance with the principle of prevention, reduction, and control of pollution coming from ships navigating through the Caspian Sea (Article 11, Para. 8f). Also, its state parties shall prevent, reduce, and control the pollution of the Caspian Sea from any source (Article 15). Acts of pollution of the marine environment are contrary to the generally recognized principles of international law of the sea (freedom of the seas, principle of protection and rational use of the living resources, principle of protection of the environment, etc.). All international agreements that require the prevention of pollution to the marine environment from any source, in fact, integrate and develop the principle of protection of the marine environment. Nowadays, maritime protection against pollution is covered in both general multilateral[27] and regional[28] conventions. The provisions of the Tehran Convention, as presented below, refer to a number of such conventions and their provisions, especially the 1982 United Nations Convention on the Law of the Sea, which reflects the rules of customary international law, as well as the 1997 Convention on the Law of the Non-navigational Uses of International Watercourses. The Tehran Convention defines pollution as "the introduction by man, directly or indirectly, of substances or energy into the environment resulting or likely to result in such deleterious effects as harm to living resources and marine life, hazards to human health and hindrance to legitimate uses of the Caspian Sea,", almost literally echoing the UNCLOS provisions.[29] Existing international environmental law defines the states' duties and rights within the field of environmental protection and provides for the most comprehensive general framework covering all forms and sources of pollution. Similarly, regulation of a variety of sources of pollution is included in the Tehran Convention, including pollution from land-based sources, seabed activities, vessels, and other human activities, as well as pollution by dumping.

The Caspian Sea Convention also regulates the issue of pollution. It defines "pollution" as *the introduction by man, directly or indirectly, of substances, organisms or energy into the ecological system of the Caspian Sea, including from land-based sources, which results or is likely to result in deleterious effects, such as harm to biological resources and marine life, hazards to human health, hindrance to marine activities, including harvesting of aquatic biological resources and other legitimate uses of the sea, impairment of quality for use of sea water and reduction of amenities* (Article 1).

10.4.1 Land-Based Pollution

The prevention and elimination of pollutants and any adverse impact of human activities upon the maritime environment is mostly related to land-based sources of pollution, such as waste and wastewater from specific household, industrial, and agricultural activities. The pollutants enter the maritime environment through dumping or incineration of waste or other items on land.

---

[27]International Convention for the prevention of pollution of the sea by oil of 1954, Convention on the High Seas of 1958 (Article 24, 25): UNCLOS; London Convention 1972, MARPOL.

[28]UNEP Regional Seas Programme; OSPAR Convention 1992; Baltic Convention (1992).

[29]Tehran Convention Article 1; UNCLOS, Article 1(4).

One of the considerable sources of land-based pollution is agricultural production. Agricultural pollution involves the loss of soil and use of chemicals, which contaminate groundwater and rivers and, eventually, the seas. Pollution emerges also from industry and advanced technology. Pollution from land-based sources includes groundwater and river pollution, which eventually enters the marine environment, as well as airborne pollution, which, through the atmosphere, reaches the ocean.

The Caspian Sea is a closed drainage system, so entering pollutants can hardly be removed. The main sources of pollution, inflowing via rivers Volga, Ural, and Kura, are agriculture, industry, and urbanization.[30] Agricultural pollutants (environmentally harmful pesticides) come mainly from small-scale farms along the Caspian coastline. Newly established farms are dependent on the large-scale use of pesticides, as well as irrigation, to ensure adequate production. Industrial discharges originate mainly from wastewater treatment plants. Oil pollution affects especially the Absheron Peninsula in Azerbaijan, the waters outside Hazar in Turkmenistan, and Atyrau in Kazakhstan.

Land-based pollution is defined by the Tehran Convention as follows: "... pollution of the sea from all kinds of point and non-point sources based on land reaching the marine environment whether water-borne, air-borne or directly from coast, or as a result of any disposal of pollutants from land to sea by way of tunnel, pipeline or other means" (Article 1). This wording is an almost literal restatement of the analogous provisions of the 1974 Convention for the Prevention of Marine Pollution from Land-Based Sources. Following Article 207 of UNCLOS, the Tehran Convention requires contracting parties to "take all appropriate measures to prevent, reduce and control pollution of the Caspian Sea from land-based sources" (Article 7.1).

Similarly to the provisions of some of the UNEP Regional Seas Protocols,[31] the Tehran Convention requires state cooperation where watercourses flow through the territories of two or more countries. In the case of discharge likely to cause pollution to the Caspian Sea, states "shall co-operate in taking all appropriate measures ... including, where appropriate, the establishment of joint bodies responsible for identifying and resolving potential pollution problems" (Article 7.3). Further, the Tehran Convention sets emission standards, which are additional measures of prevention, reduction, and control of pollution in the Caspian Sea, using the best

---

[30]Caspian Sea state of environment 2011, Report by the interim Secretariat of the Tehran Convention for the Protection of the Marine Environment of the Caspian Sea and the Project Coordination Management Unit of the "CaspEco" project, pp. 28–32, See: http://www.grida.no/publications/caspian-sea/, Accessed 1 July 2014. Section on land-based pollutants was based on the first and the second Transboundary Diagnostic Analyses, Rapid Assessment of Pollution Sources studies performed by all littoral states (2007), the Baseline Inventory Report: Land- based point and non-point pollution sources in the Caspian Coastal Zone (2008) and the Regional Pollution Action Plan (2009).

[31]Athens LBS Protocol 1980; Quito LBS Protocol 1983; Kuwait LBS Protocol 1990.

available environmentally sound technology, best environmental practice, and low- and nonwaste technology.[32]

The Moscow Protocol, which, however, is not yet in force, is to be seen as a means of implementing the Tehran Convention's (Article 7) requirement to prevent, reduce, and control pollution from land-based sources, in particular, pollution of the sea from all kinds of point and diffuse sources based on land reaching the marine environment, whether waterborne, airborne, or directly from the coast. The Moscow Protocol applies, firstly, to emissions of polluting substances originating from land-based points and to diffuse sources that have or may have an adverse effect on the marine environment and/or coastal areas of the Caspian Sea; secondly, to inputs of polluting substances transported through the atmosphere into the marine environment of the Caspian Sea from land-based sources under the conditions defined in Annex III; and, thirdly, to pollution resulting from activities that affect the marine environment and/or coastal areas of the Caspian Sea.

In implementing the Moscow Protocol, the contracting parties should (Article 5), firstly, adopt regional and/or national programs or plans of actions based on pollution source control and containing measures and, where appropriate, timetables for their completion and, secondly, address 12 different activities (agriculture and animal husbandry, industry, water, waste management, tourism, etc.)[33] and 15 substances[34] through the following:

- Progressive development, adoption, and implementation of emission controls, including emission limit values for relevant substances, environmental quality standards, and environmental quality objectives, as well as management practices based on the factors defined in Annex I; and

---

[32]Tehran Convention, Article 7.2; see also Montreal Guidelines and OSPAR Convention 1992.

[33]1. Agriculture and animal husbandry; 2. Industry (Aquaculture; Electronic; Energy production; Fertilizer production; Food processing; Forestry; Nuclear; Metal industry; Mining; Oil and gas related activities; Paper and pulp; Pharmaceutical; Production of construction materials; Production and formulation of biocides; Recycling;·Shipbuilding and repairing; Tanning; Textile; Waste management: Hazardous and toxic waste; Industrial Wastewaters; Municipal solid waste and wastewaters; Radioactive waste; Sewage sludge disposal; Waste incineration and management of its residues; Rocket fuel; 4. Tourism; 5. Transport; 6. Construction and management of artificial islands; 7. Construction of motorways and highways; 8. Liquidation of chemical weapons and ammunition; 9. Dredging; 10. Construction of harbours and harbour operations; 11. Alteration of the natural physical state of the coastline; 12. Installations out of exploitation which are affected by sea-level fluctuations.

[34]1. Bioaccumulation and biomagnification. 2. Cumulative effects of substances. 3. Distribution patterns of substances (i.e. quantities involved, use patterns and probability of reaching the marine environment). 4. Effects on the organoleptic characteristics of marine products intended for human consumption. 5. Effects on the smell, color, transparency, temperature or other characteristics of seawater. 6. Health effects and risks. 7. Negative impacts on marine life and the sustainable use of living resources or another legitimate uses of the sea. 8. Persistence of substances. 9. Potential for causing eutrophication. 10. Radioactivity. 11. Ratio between observed concentrations and no observed effect concentrations (NOEC). 12. Risk of undesirable changes in the marine ecosystem and irreversibility or durability of effects. 13. Toxicity or other noxious properties (e.g. carcinogenicity, mutagenicity, teratogenicity). 14. Capability of long-distance transport.

- Including timetables for achieving the limits, management practices, and measures agreed by the states; ... [.]

States should also utilize or promote best available technologies (BAT), best environmental practices (BEP), and the transfer of environmentally sound technology (Annex V). Further, Caspian states should (Article 6) progressively formulate and adopt common guidelines as well as regional programs and plans of action, dealing in particular with, first, the length, depth, and position of pipelines for coastal outfalls by considering, in particular, the methods used for the treatment of emissions; second, special requirements for emissions calling for a separate treatment; third, the quality of seawater that is necessary for the protection of human health, living resources, and ecosystems when used for specific purposes; fourth, the control and, where necessary, progressive replacement of products, installations, and industrial and other processes causing significant pollution to the marine environment and coastal areas; and, fifth, specific requirements concerning the quantities of the substances discharged, listed in Annex I to the Protocol; their concentration in emissions; and the methods of discharging them.

The Caspian states should ensure (Articles 7–9) that the emission controls of the point sources of pollution and the methods of control of diffuse sources of pollution, especially from agricultural activities, as well as other activities not mentioned in Annex I, which may have adverse effect on the maritime environment or coastal areas, are based on BAT and BEP. States are required to take all appropriate measures, including national action plans and states' limitations on point sources, to reduce inputs of pollutants.

Regional measures for preparedness, response, and cooperation for the protection of the Caspian Sea from oil pollution from land-based sources are regulated in the Aktau Protocol Concerning Regional Preparedness, Response and Co-operation in Combating Oil Pollution Incidents to the Tehran Convention on the Protection of the Marine Environment of the Caspian Sea adopted in 2011 but not effective yet. It was signed in addition to the Tehran Convention by all Caspian states on August 12, 2011, and entered into force on July 25, 2016. This Protocol covers operational measures (Article 8) related to oil spills from ships, offshore units, seaports, and oil-handling facilities (Article 9):

> The objective of this Protocol is to provide regional measures for preparedness, response and co-operation for protection of the Caspian Sea from oil pollution caused by activities referred to under Articles 8 and 9 of the Convention and marine oil pollution originating from land-based sources (Article 3).

The Protocol requires states (Article 4) to jointly develop and establish guidelines for the practical, operational, and technical aspects of joint action, as well as a regional mechanism. Its operational implementation should be based on a Caspian Sea plan concerning regional cooperation in combating oil pollution in cases of emergency, which was originally drafted during the consultations of the Caspian littoral states (2001–2003), working in close cooperation with the Caspian Environment Programme and the International Maritime Organization. Further, the Protocol requires states to establish national systems and contingency plans for combating oil pollution incidents (Article 5) and pollution reporting procedures (Article 7), enabling them to take operational measures in case of oil spill. States should take the necessary measures to ensure that ships flying their flag carry on board a shipboard oil pollution emergency plan. A contracting party requiring assistance to deal with an oil pollution incident, or the threat of such an incident, may request assistance, upon reimbursement of costs of assistance, from the other contracting parties (Article 10).

10.4.2 Pollution from Seabed Activities

The main reason for pollution from seabed activities is the escape of harmful substances emerging from the exploitation, exploration, and processing of raw materials on the seabed, primarily through oil and gas drilling. Leaked oil forms a film, spreading on the water surface, which interrupts the interaction between water and the atmosphere, fouls the feathers of seabirds, and pollutes flora and fauna. Even soil gets strongly saturated. The extraction of nonliving resources in the Caspian Sea is significant even on a worldwide scale. Caspian oil and gas industry is developing especially in Azerbaijan, Kazakhstan, and Turkmenistan. As a result, ecological degradation reaches a significant level. Offshore oil production, faulty pipes, unavoidable accidents, and effluents from refineries and the petrochemical industry in the Caspian cause extensive environmental damage.

The Tehran Convention gives the littoral states of the Caspian Sea the mandate "to prevent, control and reduce pollution of the Caspian Sea resulting from seabed activities" (Article 8). The Caspian states' commitment to merely take all appropriate measures to reach the set goal is weaker than that under UNCLOS (Article 208) and some regional agreements,[35] which require parties to adopt laws and regulations as well as other measures to prevent, reduce, and control pollution. The Tehran Convention merely encourages parties to cooperate in the development of protocols to that Convention to prevent, control, and reduce the pollution of the Caspian Sea caused by seabed activities.

Regional measures for preparedness, response and cooperation for the protection of the Caspian Sea from oil pollution caused by seabed activities, and also from vessels and land-based sources, are regulated by the Aktau Protocol to the Tehran Convention. See for more details Sect. 10.4.1.

---

[35]OSPAR Convention 1992, Article 5 and Annex III; UNEP Guidelines on offshore Mining and Drilling 1982, Baltic Sea Convention (1992), Article 12(1).

10.4.3 Pollution from Other Human Activities

The variety of human-induced negative impact on the marine environment cannot be limited to the problem of pollution only, although pollution is classified as the largest, most common and therefore the most dangerous factor. Numerous other factors of anthropogenic impact on the marine environment include changes in temperature and radioactivity, introduction of wastewater and influx of nutrients, irretrievable water use and destruction of aquatic organisms by seismic surveys, cultivation of arable species, destruction of coast, construction of oil platforms, etc. Many of the purely land-based activities, such as dam construction, installation of irrigation constructions, deforestation, or atmospheric fumes, can cause a negative impact even a hundred kilometers away from the coast. The coastal population is growing, and their increased activity in the coastal zone essentially changes the local environment. Clearing, land reclamation, building of drainage for flood protection, construction of roads and ports, etc. often accelerate coastal erosion and destroy the habitat.

In its Article 11, the Tehran Convention requires states "to take all appropriate measures to prevent, reduce and control pollution of the Caspian Sea resulting from other human activities ... including land reclamation and associated coastal dredging and the construction of dams." Their consequences for the Caspian environment and economy (especially the fishing industry) are harmful.[36]

The ecosystem of the Caspian Sea is highly negatively impacted by the extending land reclamation activities. This human interference caused the spreading of steppe and desert species in recent years, which are typical of zonal communities that are poor in species and are characterized by low productivity. This causes the disappearance of rare and endemic species. A historical example supporting this thesis in the Caspian region is the disappearance of almost the entire population of the Caspian tiger in the early twentieth century. Here, as elsewhere, socioeconomic necessity collides with ecological needs. It is visible in the case of a development program for the Kazakh sector of the Caspian Sea. This program aims to increase offshore oil production by 2015 and thus enforces the development of the necessary infrastructure in the country—and thus land reclamation.

Additional untreated sewage and pollutants are discharged into the Caspian Sea from the catchment area of the Volga, more than 1 million km^2, and from the Kura, as well as the rivers of Ural. Another environmentally harmful human activity is dredging, which on the one hand contributes significantly to coastal erosion but on the other hand responds to economic necessity. This conflict of interests becomes visible in the case of the Volga–Don waterway. Due to a deficient road system in the region the river course serves as essential mean of transporting goods, but for this reason there is frequent need for invading into its natural system, and for instance to deepen it to fulfil the transportation needs. As the number of tankers in operation is not sufficient, there is a need to use larger vessels in the Volga–Don Canal to

---

[36]See: (Barannik et al. 2004) et al., pp. 45 et seq.

transport oil to the Azov Sea and the Black Sea. It requires, however, dredging of the Volga–Don Canal, which is filled with a lot of silt.[37]

Another environmental problem indicated by the Tehran Convention is pollution originating from the construction of dams. The land and the water are ecologically intertwined. Any interruption of this connection, as is the case when a dam is constructed, causes extensive changes to the river system and its hydrology and natural flow. A cascade of reservoirs built on the Volga improves water supply and inland water transport, but at the same time the increased demand for water from the industry and settlements, as well as the evaporation of artificial reservoirs, leads to larger water losses and thus to the increase of salinity of water. The dams and locks on the Volga raise the water level in the river, thus reducing the flow rate, so that the self-cleaning ability of the Volga gets limited. Polluted waters cannot be purified in a natural way as the respective biological species are covered by sediments, which do not get flushed out because of low water flow rate. The most serious example of damage to the environment of the Caspian Sea is the Kara–Bogaz–Gol Bay, which was separated from the Caspian Sea by a dam. The dam existed from 1980 to 1992 and was built to prevent water runoff from the Caspian Sea and the falling of water level. Instead, it had disastrous ecological consequences. The sturgeon was separated from its spawning areas, the bay turned into a salty lagoon, salt storms ravaged the coast, coast desertification was accelerated, etc.

10.4.4 Pollution by Dumping

In the 1970s of the twentieth century, the scope of environmental law was extended. Instead of referring merely to the prevention of pollution of the marine environment from oil, for the first time it addressed the problem of pollution from other sources[38] and from dumping. Pollution by dumping refers to pollution that is created on land and subsequently transported for disposal at sea. The pollution of the sea by solid waste of all kinds that either deliberately or accidentally enters the oceans each year is caused by waste and also by washed-out agricultural soil, pesticides, other chemicals, and effluents that are washed into the water cycle. The chemical industry sinks huge amounts of waste in the sea, of which the dumping of heavy-metal-contaminated dredged material is particularly problematic. The dumping of radioactive waste is particularly dangerous and forbidden under international law. Many pollutants get into the groundwater and eventually also into the oceans.

Many media sources state that the Caspian Sea suffers from dumping from different sources. Naval ships regularly dump their ballast water into the Caspian Sea. Untreated industrial wastewater, including nuclear, is dumped into the Caspian annually. Another point of concern is the pollution of the Caspian Sea originating

---

[37] See: Gouseinli.
[38] Intervention Protocol of 1973.

from onshore and offshore oil. Already at the time of the Soviets, it was a few times higher than the maximum permissible concentration.[39] Also, nowadays, oil waste is damped from platforms in the Caspian Sea. The most endangered areas are Baku Bay, Absheron archipelago, Turkmenbashi, Cheleken, Mangishlag, Tengiz, etc. Offshore drilling products and wastewater from cleaning facilities are often dumped into the Caspian Sea. Flooded oil wells due to sea level rise pose increasing environmental danger. In the Kazakh sector of the Caspian Sea alone there are 19 oilfields with 1485 oil wells in the coastal zone. Having been constructed many years ago, the structure of offshore oil wells, fields, and pipelines has started to corrode, thus polluting the sea.[40] A broad set of international principles established for combating pollution caused by dumping are included in UNCLOS (Article 210) and the 1972 Convention on the Prevention of Marine Pollution by Dumping of Wastes and Other Matter (further referred to as London Dumping Convention). The London Dumping Convention, for instance, includes measures, procedures, and standards aimed at the prevention, reduction, and control of pollution from vessels. This approach, visible in a number of international agreements, among others the UNEP Regional Seas Protocols,[41] reflects the current state of international standards relating to the prevention of pollution, including marine pollution by dumping. These principles are confirmed in contemporary case law.[42]

As it includes merely a general reference to existing sources of international law, the Tehran Convention does not expressly reflect the obligation to refer to dumping protection mechanisms existing in law. Following the London Dumping Convention, the Tehran Convention characterizes dumping as "any pollution to the Sea from any deliberate disposal into the marine environment of waste or other matter from vessels, aircraft, platforms, or other man-made structures in the Caspian Sea or any deliberate disposal of vessels, aircraft, platforms, or other man-made structures in the Caspian Sea" (Article 1). Also, the general requirements regarding the prevention, reduction and control of pollution are common for both treaties on the global level and those for the Caspian region.[43]

However, the Tehran Convention's provisions on dumping are more general. They do not mention any particular types of pollutants, as opposed to the 1972 London Dumping Convention, which established three categories of waste, clearly prohibiting dumping only in the case of highly hazardous waste substances listed in Annex I, except in emergency cases.[44]

---

[39]Hekimoğlu (1999).

[40]Alexander Holstein et al. (2018).

[41]Barcelona Dumping Protocol 1976, Article 2; Noumea Dumping Protocol 1986, Article 2; Paipa Dumping Protocol 1989, Article 1.

[42]Case concerning Trail Smelter Arbitration (USA v. Canada), 3. UNRIAA 1905 (1941); Corfu Chanel Case (1949) ICJ 4; Lake Lanoux Arbitration (Spain v. France), 12 UNRIAA 281 (1963).

[43]UNCLOS, Article 1.1(5), 210.1; Tehran Convention, Article 10.1.

[44]London Dumping Convention, Article V; Interim Procedures and Criteria for Determining Emergency Situations, LDC V/12, Annex 5.

Dumping of other substances requires a "special" or "general" permit, applied for in advance, which could also be applicable in the Caspian Sea case.[45] The Tehran Convention does not prejudice the question of at which threshold pollution from dumping becomes impermissible. According to the London Dumping Convention, national authorities should keep detailed records of all relevant matters (the characteristic and composition of the substance and of the dumping site, the methods of deposit, and other general considerations and conditions such as the possible effects on marine life, other uses of the sea, and the practical availability of alternative methods of treatments, disposal, or delimitation).[46] Compared to the provisions of the London Dumping Convention, the Tehran Convention limits the jurisdiction of the states parties to ships and air vehicles that are registered in their territory or fly their flag. The only exception to the rules of the Tehran Convention for preventing, hindering, reducing, and controlling dumping in the Caspian Sea is allowed in emergency situations. Such an emergency case applies only to situations where human or marine life is threatened or aircraft or vessels are in danger of "complete destruction or total loss" and must be reported to the contracting parties. The dumping is allowed in case if it is "the only way of averting the threat, and if there is every probability that the damage consequent upon such dumpingwill be less than would otherwise occur. Such dumping should be conducted as to minimize the likelihood of damage to human or marine life or hindrance to legitimate uses of the sea in accordance with the applicable international and regional legal instruments" (Article 10 Para. 3).

Also the Ashgabat Protocol to the Tehran Convention obliges states to strengthen the regulation "pertaining to the release or dumping of wastes and other substances likely directly or indirectly to impair the integrity of the area" (Annex 2, C 4(a)).

10.4.5 Pollution from Vessels

Vessel-source pollution is generated by marine transportation, which includes international operational and accidental discharges resulting from shipping operations. Vessel-source pollution includes all types of pollution originating from vessels, such as oil, chemicals, natural gas, and other hazardous materials, which result from accidents at sea, reballasting, and tank cleaning. An example of a significant increase in the pollution of the Caspian Sea associated with oil transportation by tankers was the increase in the amount of oil transported from terminals in Turkmenbashi via the

---

[45]London Dumping Convention, Article IV(1)(b) and (c).

[46]London Dumping Convention, Article VI(3) and Annex III, as mentioned in 1989, Res LDC 32 (11), Amendments to the Guidelines for the Application of Annex III (LDC 11/14, Annex IV).

Caspian Sea to Baku, which was then used to fill the Baku–Tbilisi–Ceyhan pipeline built in 2005.[47]

Serious pollution accidents in the 1960s involving sunken oil tankers such as the Torrey Canyon supertanker Amoco Cadiz, the Exxon Valdez, or the Mega Borg revealed the immense danger brought by ship operations to the marine environment and thus led to increased legislative activity by the global society. To reduce the pollution of marine environment originating from ships, many international agreements were adopted. Regional standards for the prevention, reduction, and control of pollution of the marine environment by ships can be found in numerous international agreements.[48] International legal instruments that contain the most important provisions against pollution from vessels include UNCLOS and the International Convention for the Prevention of Pollution from Ships (MARPOL). UNCLOS merely empowers states to regulate pollution from vessels and limits their jurisdiction to the application of generally accepted international rules and standards contained in the relevant multilateral agreements. The most important is the MARPOL Convention, which regulates the obligation of states to prevent marine pollution. The substantive norms arising from six additional protocols to MARPOL regulate the prevention of pollution of the marine environment by oil, noxious liquid substances in bulk, harmful substances carried by sea in packaged form, sewage from ships and garbage from ships, as well as the prevention of air pollution from ships.

Mindful of this important source of pollution to which the Caspian Sea may be exposed, the parties to the Tehran Convention agreed to take all appropriate measures "to prevent, reduce and control pollution of the Caspian Sea from vessels" (Article 9). These provisions contain the basic elements found in other international instruments. The Tehran Convention requires only in a general way the cooperation of states in the development of protocols and agreements prescribing agreed measures, procedures, and standards and binds them with the relevant international rules.

Regional measures for preparedness, response, and cooperation for the protection of the Caspian Sea from oil pollution caused by vessels are provided for under the Aktau Protocol. See for more details Sect. 10.4.1.

---

[47] Geospatial Technologies and Human Rights Project—Satellite Imagery Analysis for Environmental Monitoring: Turkmenbashi, Turkmenistan May 2013, prepared by American Association for the Advancement of Science (assessment of two towns located near Turkmenbashi on the Caspian Sea: Avaza and Tarta). For full report: http://www.aaas.org/news/releases/2013/media/AAAS_Turkmenistan_Oil_2013.pdf. accessed July 18, 2020.

[48] Copenhagen Agreement 1971; Helsinki Convention 1992, Article 7, Annex IV; 1976 Barcelona Convention, Article 6; Kuwait Convention 1978, Article IV; Abidjan Convention 1981, Article 5; Lima Convention 1981, Article 4 (b); Lima Convention 1981; Jeddah Convention 1982: Cartagena Convention 1983, Article 5; Bonn Agreement 1983; Nairobi Convention 1985, Article 5; Noumea Convention 1986, Article 6.

10.4.6 Environmental Emergencies

Situations that lead to a catastrophic pollution of marine environment are often caused by unexpected and unforeseen events at sea. Therefore, there has been a need for increased contractual activities to counteract the causes of such pollution for a long time. The international agreements providing for the prevention of pollution of the sea, which were one by one concluded until the 1990s of the twentieth century, were unsatisfying because they did not require states to take preventive measures. In 1990, the first International Convention on Oil Pollution Preparedness, Response, and Co-operation (OPRC) was adopted, giving necessary impetus for subsequent international legal instruments. The agreement contains a general obligation of states parties to take all appropriate measures to prepare for oil pollution incidents and to fight them.

The oil disaster caused by a blowout on a BP rig in the Caspian Sea, off the coast of Baku, in September 2008, was the most serious example of oil pollution on the Caspian.[49] There were also other examples of unpredictable pollution to the Caspian marine environment caused by humans or nature because the Caspian Sea is one of the main transport routes in the region. Recognizing that it is desirable to protect the fragile environment of the Caspian Sea from environmental emergencies, the Tehran Convention devotes a separate article to this problem.

Defining environmental emergency, the Tehran Convention refers to the basic elements contained in provisions regarding international watercourses,[50] saying that an environmental emergency is "a situation that causes damage or poses an imminent threat of pollution or other harm to the marine environment of the Caspian Sea and that results from natural or man-made disasters" (Article 2). Further, states are required to take all appropriate measures and cooperate to protect human beings and the marine environment using preventive, preparedness, and response measures against the harm caused by natural or man-made emergencies (Article 13). This provision refers to the obligations contained in the OPRC Convention, as well as the subsequent Protocol on Preparedness, Response and Co-operation to Pollution Incidents by Hazardous and Noxious Substances of 2000. Both are aimed at providing a global framework for international cooperation in combating major incidents or threats of marine pollution.

Following OPRC Convention principles, which provide for national and regional systems for preparedness and response (Article 6), the Tehran Convention requires states to "take all appropriate measures to establish and maintain adequate emergency preparedness measures, including measures to ensure that adequate equipment and qualified personnel are readily available to respond to environmental emergencies" (Article 13.4). The Tehran Convention, like the OPRC Convention (Articles 5 and 7), promotes international cooperation in industrial accidents and environmental emergencies. It obliges states to set up an early warning system and to

---

[49]Palast (2012).
[50]UN Water Convention, Article 28.

prescribe necessary actions in the event of an environmental emergency or imminent threat. "The Contracting Party of origin should ensure that the Contracting Parties likely to be affected, are, without delay, notified at appropriate levels" (Article 13.3). For the coordination of the means of communication and the promotion of cooperation, it provides a number of legal instruments concluded under the auspices of UNEP.[51] According to the abovementioned state obligation to undertake preventive measures and set up preparedness measures, every state should first identify hazardous activities within its jurisdiction that are likely to cause environmental emergencies. The Caspian littoral states are not expressly committed to developing specific national and regional systems and contingency plans for the prevention and combat of pollution accidents, which is not the case under the OPRC Convention (Article 6). The only aspect that was taken from the OPRC Convention is the obligation to design a competent national authority that is responsible for the prevention and combating of oil pollution (Article 13(4)). After that, the Tehran Convention requires member states to ensure that other contracting parties are notified of any such proposed or existing activities. "The contracting parties shall agree to carry out environmental impact assessment of hazardous activities and to implement risk-reducing measures" (Article 13.2). It will be referred to in detail in the following chapter. Damage to the environment resulting from an emergency situation is regulated in the Agreement on cooperation in the field of prevention and liquidation of emergency situations in the Caspian Sea, which was adopted on September 29, 2014 (more details in Sect. 4.5.3).

10.5 Protection, Preservation, and Restoration of the Marine Environment

The marine environment of the Caspian Sea—including its water and the adjacent coastal areas—represents a self-contained unit, being an indispensable component of a life-supporting system, that requires sustainable development. The Tehran Convention includes the general obligation of states to individually or jointly take all appropriate measures to protect, preserve, and restore the environment of the Caspian Sea (Article 4.b). The provision is included in the international instruments regarding both seas[52] and international watercourses,[53] as well as others.[54] The term "environment" is to be interpreted quite broadly, to be applied to areas "surrounding" the Caspian Sea, which have minimal influence on the protection and

---

[51]Mediterranean Emergency Protocol (1976), Article 1; Kuwait Emergency Protocol 1978, Article II; Jeddah Pollution Emergency Protocol 1982, Article II.

[52]UNCLOS, Article 192.1.

[53]UN Water Convention, Article 20.

[54]African Convention on the Conservation of Nature and Natural Resources 1968, Article II; ASEAN Agreement 1985, Article 1.

10.5 Protection, Preservation, and Restoration of the Marine Environment

preservation of the Caspian Sea itself. It is derived from the provisions obliging states to take necessary measures to develop and implement national strategies and plans for the planning and management of land affected due to its proximity to the sea (Article 15). The duty to protect the environment requires states to shield it from damage or harm, as well as from significant threats of harm, which reflects the general application of the principle of precautionary action. The protection, as well as the preservation, should ensure its continued viability as a life-support system, providing a basis for sustainable development. The requirement that states should act individually or jointly is to be understood to mean that appropriate action is to be taken where necessary and on an equitable basis.

In view of the general nature of the obligation contained in the abovementioned article, it is preceded by other more specific articles in Part IV. The Tehran Convention requires particular regard from the contracting parties for the protection, preservation, restoration, and rational use of marine living resources (Article 14). The ocean resources are usually categorized as nonliving and living, where the latter include fish stocks and marine mammals. The goals of international law for fishery conservation are as follows: promoting international cooperation, managing and conserving fisheries and marine living resources, and supporting international research, scientific cooperation, and international regulation.

Citing the need to apply the best scientific evidence available, the Tehran Convention contains provisions requiring parties to the Convention to take all appropriate measures required to protect, preserve, and restore the marine environment. In doing so, the Convention mostly repeats objectives included in the United Nations Conference on Environment and Development (UNCED) Agenda 21.[55] The requirement of the Tehran Convention (Article 14.1) and UNCED (Article 17.46(a)) refers to the development of living resources' potential for conservation, restoration, and rational use of environmental equilibrium to satisfy human needs in nutrition, as well as meeting social and economic objectives. The provisions on the conservation and management of fisheries may be considered as reflecting customary international law, which is the reason that the Tehran Convention almost literally restates the obligation imposed on states as contained in UNCLOS (Article 61(3)). It requires the Caspian Sea littoral states to ensure the maintenance and restoration of the populations of marine species at a level that can produce the maximum sustainable yield, i.e., the largest average catch that can be taken continually (in a sustainable fashion) from a stock under conditions of relevant environmental and economic factors and relationships among species (Article 14.1b). The already-mentioned requirement is in addition to the other obligations of states, also to be found among UNCLOS provisions, which is to ensure that marine species are not endangered by overexploitation, i.e., the catches exceed the maximum sustainable limit recognized by the Food and Agriculture Organization.[56] Like UNCED, the Tehran Convention in Article 14(1)(d) provides for the promotion of the "development and

---

[55]Para. 17.46, A/CONF.151/26 (Vol. II) (1992).
[56]Tehran Convention, Article 14.1c; UNCLOS, Article 61(1) and (2).

use of selective fishing gear and practices that minimize waste in the catch of target species and that minimize by-catch of non-target species." According to UNCLOS, both of these legal instruments require states to "protect, preserve and restore endemic, rare and endangered marine species" (Article 194.5).

States parties are committed to the rational management of the Caspian resources, as well as exploration, protection and conservation on its environment (Preamble). Further on, the Convention introduces principles that shall guide the activities of the Caspian Sea states, including the protection of the environment of the Caspian Sea, as well as the conservation, restoration, and rational use of its biological resources (Article 3 and Article 4).

10.5.1 Protection of Biodiversity

The Caspian Sea's biological diversity is characterized by a high level of endemic fauna species, present in the mid-Caspian Sea region, and their great diversity, especially in the north Caspian. The total count of species in the Caspian Sea region is estimated at up to 2000 groups of plants and animals.[57] There are many algae species and more than 100 other native species but only one marine mammal—the Caspian seal. As many as 19 fish species are listed in the International Union for Conservation of Nature (IUCN) and some national Red Data Books. There are more than 300 species of birds.

The Caspian states' commitment regarding the protection, preservation, restoration, and rational use of the marine living resources in the Caspian Sea within the Tehran Convention refers to the main objective of the 1992 Convention on Biological Diversity, i.e. conservation of biodiversity. In Article 14(1)(f), the Tehran Convention requires states to pay particular attention to the "habitats of rare and endangered species, as well as vulnerable ecosystems." The Tehran Convention also requires states to protect, preserve, and restore biological resources (Article 14.2), which demand the regulation or management of biological resources important for the conservation of biological diversity, whether within or outside protected areas, with a view to ensuring their conservation and sustainable use.[58] The Tehran Convention does not include the direct definition of the biological resources. Instead, the Biodiversity Convention defines them as "genetic resources, organisms or parts thereof, populations, or any other biotic component of ecosystems with actual or potential use or value for humanity" (Article 2). Competing with the incentive for cooperation between governmental authorities and the private sector in developing methods for the sustainable use of biological resources contained in the Biodiversity Convention (Article 10e), the Tehran Convention requires states to "co-operate in the development of Protocols in order to undertake the necessary measures for

---

[57]Caspian State of Environment, pp. 54–62.
[58]Convention on Biological Diversity of 1992, Article 8b.

protection, preservation and restoration of marine biological resources" (Article 14.2).

In the Tehran Convention, the Caspian states assigned priority to the development of an ancillary Biodiversity Protocol, which was adopted in May 30, 2014. At the same date Iran, Russia and Turkmenistan signed the Protocol until now only Turkmenistan has ratified it when ratification by other contracting parties is missing. According to the Article 1 of the Protocol 'biological diversity' means "the variability among living organisms from all sources including, inter alia, terrestrial, marine and other aquatic ecosystems and the ecological complexes of which they are part; this includes diversity within species, between species and of ecosystems:"

> The objectives of this Protocol are to protect preserve and restore the health and integrity of the biological biodiversity and ecosystem of the Caspian Sea, as well as to ensure the sustainable use of biological resources (Article 2).

This Protocol applies to the maritime environment of the Caspian Sea as well as land affected due to its proximity to this sea, including wetlands of international significance (Article 3). The Protocol includes provisions regarding the protection and conservation of species and of protected areas. It includes instruments and prescribes requirements for the protection and conservation of species, including measures for the protection and conservation of species and regulation regarding alien species and genetically modified species. It provides a conservation framework for biodiversity in the coastal zones. It also defines access to genetic resources and to technology. It further defines significant technical cooperation in the system as well as exchange of information, the obligation for environmental legal assessment, a system of reporting, as well as environmental education and public awareness obligations. It includes also provisions on the protection and conservation of protected areas, defining protected areas and their management, as well as procedures for the establishment and listing of protected areas. Also, the objective of the EIA Protocol to the Tehran Convention is to promote the conservation of the Caspian Sea's biodiversity (Article 2). Moreover, the Ashgabat Protocol stipulates cooperation of states in the development of the protocols in order to undertake necessary measures for the protection, preservation, and restoration of the marine biological resources (Article 14, Para. 2).

10.5.2 Invasive Alien Species

Invasive alien species are animals and plants that are introduced, accidently or deliberately, into a natural environment where they are not normally found, causing serious negative consequences for the original species, communities, or habitats. The enormous dissemination of invasive alien species, especially in recent years, results from increased international traffic and trade. Some of the new species were deliberately introduced into the Caspian Sea, e.g., to increase the productivity of the species living there. However, it has not always achieved the desired effect, such as in the case of *Azolla pinnata*, which caused anoxia in Iran's lagoons.

Invasive alien species are considered one of the greatest threats to biodiversity. They are in competition with native species and threaten to displace them or cause introgression of their genes into native species. Invasive species can alter site conditions and thus ecological cycles. Even very small populations, by causing damage to native species, can cause economic losses, e.g. depriving people of their livelihood and thus causing poverty or increase in poaching etc. Recognition of these dangers contributed recently to the need to combat these species.[59]

But it was not until the 1970s that the scientific community began reviewing this problem in detail.[60] Numerous global[61] and regional[62] agreements were signed to prevent the negative impact of the invasive alien species on native plant and animal species. The development of international legal instruments relating to this issue had been continuing until the adoption of the International Convention for the Control and Management of Ships' Ballast Water and Sediments on February 13, 2004. Decision VI/23 of Conference of the Parties to the Convention on Biological Diversity is a catalog of measures for the development of national strategies for the implementation of international legal provisions on the combat of alien species. It defines the term "alien species" as "species, subspecies or lower taxon, introduced outside its natural past or present distribution; includes any part, gametes, seeds, eggs, or propagates of such species that might survive and subsequently reproduce."[63]

The definition of pollution contained in the Tehran Convention does not include biological alterations. The problem of alien species received special attention in the Tehran Convention. It was regulated in a separate article confirming the current approach that alien species are not to be defined as pollution per se.[64] Alien species were defined in the Tehran Convention as follows:

> An alien species whose establishment and spread may cause economic or environmental damage to the ecosystems or biological resources of the Caspian Sea (Article 1).

---

[59] Global Invasive Species Programmes, European Plant Conservation Strategy.

[60] MEPC Res (1991) 50(31); Convention on Biological Diversity of 1992, Objectives, Article 8h; Conference of the Parties to the Convention on Biological Diversity in 1998, Decision IV/5; in 2002, Decision VI/23; 2002 World Summit on Sustainable Development, Plan of Implementation, Para. 34(b); IMO, Res A.774(18) in 1993 and A.868(20) in 1997.

[61] 1991 MEPC Resolution 50(31); 1992 Convention on Biological Diversity, Objectives, Article 8h; Conference of the Parties to the Convention on Biological Diversity in: 1998-Decision IV/5, in 2002 Decision VI/23; 2002 World Summit on Sustainable Development, Plan of Implementation, Para. 34(b); IMO, Resolution A.774 (18) in 1993 and A.868 (20) in 1997.

[62] Convention on the Conservation of European Wildlife and Natural Habitats 1979.

[63] Conference of the Parties to the Convention on Biological Diversity (2002). https://www.cbd.int/cop/ accessed July 18, 2020.

[64] Report of the International Law Commission on the work of its 46th session, and next International Legal Committee, Commentary, 1994, Yearbook of the International Law Commission, vol. 2 pt. 2, p. 122.

Unlike the UN Water Convention (Article 22), which refers to the new species that are genetically altered or produced through biological engineering, the Tehran Convention refers to alien species, i.e., those that are nonnative to the Caspian basin and whose establishment and spread may cause economic or environmental damage to the ecosystem or biological resources of the Caspian Sea. The Tehran Convention addresses the prevention of the introduction, control, and combating of invasive alien species in a separate article because their introduction into the ecosystem is not generally regarded as pollution per se.[65] It is clear that alien or new species of flora and fauna can have an adverse influence on the marine environment of a particular water basin. They can destroy the ecological balance and result in serious problems, including the clogging of intakes and machinery; spoiling of recreation; acceleration of eutrophication; disruption of food webs; elimination of other, often valuable species; and transmission of diseases. The Tehran Convention requires littoral states to "take all appropriate measures to prevent the introduction into the Caspian Sea and to control and combat invasive alien species, which threaten ecosystems, habitats or species" (Article 12). This formulation contains basic elements of the UNCLOS provisions regarding the preservation, reduction, and control of introduction of alien and new species, which may cause significant and harmful changes thereto (Article 196). While any introduction of alien species should be under strict observation, the Tehran Convention does not indicate any precautionary action regarding alien species, unlike the UN Water Convention, which requires measures against species that "may" have a detrimental effect to the ecosystem (Article 22).

The Tehran Convention does not refer to any particular alien species by name. It also hardly provides for concrete means of combat that should be taken by states. Alien species have been introduced into the Caspian Sea both accidentally and intentionally, for economic purposes.[66] Some of them have caused considerable ecological disruption, such as the algae *Rhyzosolenis* becoming a dominant phytoplankton, *Rhithropanopeus harrisii* causing a complete change in the benthic area, and *Acartia tonsa* dominating zooplankton. The comb jelly (*Mnemiopsis leidyi*), which invaded in the late 1990s, competes with the tulka for food and eats its larvae, which in turn adds to the decline of the stock of Caspian seals, in whose diet tulka is a key element.

"Alien species" are defined in the Ashgabat Protocol as *any species occurring outside of its historically known natural range as a result of intentional or accidental dispersal by human activities* (Article 1). On the other hand, "invasive alien species" are defined therein as *alien species whose establishment and spread may cause economic or environmental damage to the ecosystems or biological resources or harm to human health* (Article 1).

---

[65]Report of the International Law Commission on the work of its 46th session, and next International Legal Committee, Commentary, ILCYB (1994), Vol. 2, Part 2, 122, 124.
[66]Caspian State of Environment, pp. 62–63.

10.5.3 Coastal Zone Management

The most important problems of coastal regions are coastal erosion, exacerbated by infrastructure, the construction or development of natural gas, as well as the destruction of habitats as a result of poorly planned development, land reclamation, or marine management. Problems for the sustainable management of coastal zones are also caused by the decline in biodiversity, including fish stocks, as well as the pollution of water and soil. In the case of the Caspian Sea, an additional immense threat to water management is posed by the fluctuation of water level. In many cases, the abovementioned environmental problems can cause social problems for people living there, such as unemployment, migration, competition among users, loss of development opportunities, etc.

The need for coastal zone management was highlighted for the first time in the resolution of the United Nations conference in Rio de Janeiro in 1992. In Agenda 21 of this Conference, objectives for the sustainable development of the seas and coasts were set out.[67] Many problems in the coastal zones are cross-border issues and cannot be solved by individual countries. Thus, there is a need for a supraregional state community ready to undertake appropriate cooperative policies and investment strategies. The importance of these problems is reflected in the great number of international instruments, some of them of a worldwide character,[68] providing for the development of national plans for coastal zone management.[69]

The Tehran Convention requires states to "take necessary measures to develop and implement national strategies and plans for planning and management of the land affected by proximity to the sea" (Article 15). This provision follows the 1996 Common Recommendations for Spatial Planning of the Coastal Zone in the Baltic Sea Region, recognizing coastal zone management as a broad social, economic, and ecological approach aimed at improvement activities that "influence significantly the quality of the environment, economic and social opportunities and the cultural heritage in the coastal zone."[70] Also, other international legal instruments refer to the urgent need for coastal states to develop integrated coastal zone management plans.[71]

Despite the recognition of the need for a successful coastal management, only a few initiatives have been undertaken in the Caspian region so far to remove existing

---

[67]Chapter 17 Agenda 21.

[68]Land-Ocean Interactions in the Coastal Zone, UNEP-ICAM: Integrated Coastal Area and River Basin Management, EUCC-The Coastal Union, HELCOM HABITAT Group.

[69]UNCED resolutions, Rio de Janeiro 1992, Chapter 17 of Agenda 21, statements from "The World Coast 1993" conference; Washington Declaration on the Protection of the Marine Environment from Land-Based Activities, Preamble, 23 October to 3 November 1995.

[70]Common Recommendations for Spatial Planning of the Coastal Zone in the Baltic Sea Region, Preamble (1969), Article 2b.

[71]UNCED Resolutions, Rio de Janeiro (1992), Chapter 17 of Agenda 21, statements from "The World Coast 1993" Conference; Washington Declaration on the Protection of the Marine Environment from Land-Based Activities, Preamble (23 October–3 November 1995).

obstacles. A positive exception was the preparation of a cross-border "Integrated Coastal Area Management Planning."[72] It was started under the Caspian Environmental Programme but is now outdated. The entire work should have been done by the Regional Center for Integrated Transboundary Coastal Area Management Planning (CRTC-ITCAMP) established in the Islamic Republic of Iran. It coordinated the preparation of a series of National Coastal Profiles of all Caspian littoral states, which, however, were last updated in 2003. On this basis, a "Caspian Regional Coastal Profile" for the entire region was supposed to be developed, comprising the identification of regional issues, institutional strengths and weaknesses, capabilities, training needs, and training opportunities for the Caspian region.

Reference for an integrated approach to coastal area development, based inter alia on coastal area planning, includes also the Moscow Protocol to the Tehran Convention (Article 10). It requires states to adopt and implement mitigation measures to reduce the negative impact of natural hazards, such as long-term sea-level fluctuation, storm surges, storms, earthquakes, and coastal erosion on the population and infrastructure of the coastal areas. Special attention was devoted to deforestation and land degradation, where appropriate national plans of action should contribute to combat land-based pollution of the Caspian maritime environment.

According to the Ashgabat Protocol to the Tehran Convention, the contracting parties shall take *measures to develop and adopt national coastal area management strategies. Requirement for application of these national strategy and plan include the following: (a) Integration of biological diversity conservation and environmental protection provisions in national and/or regional development planning; (b) Implementation of integrated management approach that allows multiple uses of biological resources in accordance with Article 3 of this Protocol; (c) Analysis of natural dynamics of coastal ecosystems connected with sea-level fluctuations* (Article 12).

Also, the Moscow Protocol to the Tehran Convention foresees a need for individually and, where appropriate, jointly applying an integrated approach to the development of the coastal areas (Article 10).

10.5.4 Fluctuation of the Caspian Sea Level

The Tehran Convention pays special attention to human activities intended to counteract the fluctuations of the sea level. The water level in the Caspian Sea has been rising since 1978, which has serious consequences for the entire region. The

---

[72] http://www.coastalguide.org/icm/caspian/index.html accessed July 15, 2020.

rising water level accelerates changes in the water regime, the hydrochemical regime of estuaries, and the structures and productivity of biological communities living there, as well as the chemical composition of groundwater etc. The earthen walls, which were built to isolate oil-polluted water from clean sea water (e.g., the Tengiz oil field), are vulnerable to flooding. Also, large parts of the oil infrastructure are threatened by water fluctuation. Residential areas in the coastal zone have been severely affected by the rising water level so that a resettlement of the population was needed (for example, the city of Dervish in Turkmenistan).

Environmental policy, in its pursuit of sustainable development in the areas adjacent to the Caspian Sea, depends on four major factors, among them the hydrometeorological regime of the Sea and its basin, including the fluctuations of its water level. To mitigate the economic and ecological negative consequences of the sea-level fluctuations, it is necessary to use sea-level forecasting and undertake measures that decrease sea-level variation. The International Lake Environment Committee (ILEC) Survey of the State of the World's Lakes concludes that climatic changes are the most probable cause of the fluctuations of the Caspian Sea level.[73]

The Tehran Convention deals with the issue of Caspian Sea level fluctuation, providing for states' cooperation in the development of protocols to the Convention and obliging them to carry out scientific research and jointly develop measures and procedures, in so far as is practicable, to decrease the consequences of the Caspian Sea level fluctuations:

> to co-operate in the development of Protocols to the Convention prescribing to undertake the necessary scientific research and, insofar as is practicable, the agreed measures and procedures to alleviate implications of the sea level fluctuations of the Caspian Sea (Article 16).

Further on, also few protocols to the Tehran Convention refer to the fluctuation of the Caspian Sea. The fluctuation of the Caspian Sea shall be taken into account while implementing the EIA Protocol (Article 3). Also, in the Ashgabat Protocol, it is provided that analysis of the natural dynamics of the coastal ecosystem connecting with sea-level fluctuations shall be considered while developing national strategies and plans to provide a mechanism for biological diversity conservation (Article 12). The Moscow Protocol stipulates that states shall adopt and implement mitigation measures to reduce the negative impact of natural hazards, such as long-term sea-level, fluctuations while developing coastal areas (Article 10). The measures taken to mitigate the negative consequences of Caspian Sea fluctuation include the establishment in 1998 of the so-called Caspian Centre for Water Level Fluctuations (CCWLF). It was created as part of the Technical Assistance for the Commonwealth of Independent States (TACIS) program of the European Union to support the first phase of the work of the Caspian Environmental Programme. Since 2003, the research of CCWLF has been partially, but not sufficiently, continued as part of other separate projects.

---

[73]Golubev (1997).

10.6 Institutional Framework for Cooperation in the Legal Protection of the Caspian Environment

The success of a treaty compliance regime, which constitutes a formalized monitoring of the fulfillment of the contractual obligations of states by a collective body, depends largely on how the institutional questions of competences of supervisory organs are defined in the treaty itself. Another crucial factor is that the supervisory organ should have certain instruments that would allow the sufficient verification of the accuracy of information requested by the contracting states (reports and declarations). The Tehran Convention provides for two supervisory bodies—the Party Conference and the Secretariat—which also serve as framework for institutional arrangements for the additional protocols to the Tehran Convention.

10.6.1 Conference of the Parties

The main supervisory body is the Conference of the Parties, regulated under Article 22 of the Tehran Convention. The Conference of the Parties must be appointed not later than 12 months after the entry into force of the Tehran Convention. Thereafter, the Conference of the Parties shall hold ordinary meetings at regular intervals. Extraordinary meetings of the Conference of the Parties shall be held at such other times as may be deemed necessary by the Conference of the Parties or at the written request of any Party, provided that it is supported by at least two other contracting parties. The Conference of the Parties is composed of representatives of each state party, each of which has one vote. All decisions are to be taken unanimously. The chairmanship of the Conference of the Parties shall be held in turn by each contracting party. The functions of the Conference of the Parties shall be, among others (Article 22, Section 10), to keep under review the implementation and content of this Convention, its protocols, and the Action Plan, as well as reports prepared by the Secretariat and to consider and adopt any additional protocols or any amendments. The Conference of the Parties receives and considers reports submitted by the contracting parties and reviews and evaluates the state of the marine environment and, in particular, the state of pollution and its effects, based on reports provided by the contracting parties and by any competent international or regional organization.

The Tehran Convention does not provide for "non-compliance" procedures for defaulting states. Neither does it provide for a procedure determining that a state party fails to obey it. The only aspect that the Conference of the Parties has authority to review and evaluate is the state of pollution of the Caspian Sea. However, the treaty does not specifically define the controlling procedures of compliance with the Tehran Convention and its protocols. Neither does the Tehran Convention give the Conference of the Parties the competence to ask questions to the reportable state. It significantly hinders its *de facto* control over states and the legal evaluation of the implementation of the Tehran Convention. The following provision, which regulates the sources of information available to the Conference, limits them to competent international or regional organizations. Respectively, it limits the range of potential sources of information necessary for the effectiveness of the compliance control. In terms of reporting, the parties are merely obliged to submit to the Secretariat reports on measures adopted for the implementation of the provisions of this Convention and its protocols (Article 27). The Conference of the Parties has the right merely to consider these reports.

It is to be noted that the parties deliberately left the issue of compliance control procedures of the Tehran Convention for subsequent regulation because of insufficient protection of the rights of the involved states. Aware of the need to strengthen the control mechanisms, the states agreed to cooperate in their development. It is necessary to define the means allowed to be taken following the identification of problems with compliance. Shall the supervisory body be empowered merely to make recommendations[74] or also to penalize[75] such states' behavior? If needed, the Conference of the Parties may seek technical and financial services of relevant international bodies and scientific institutions (Para. 10f) or establish such subsidiary bodies as may be deemed necessary (Para. 10g) for the implementation of this Convention and its protocols, also for the control over their fulfillment.

The Conference of Parties serves as the institutional framework for the Aktau Protocol, which is the only additional protocol to the Tehran Convention already in force. The Conference of Parties, according to Article 12, shall, inter alia, *(a) Keep under review the implementation of this Protocol; (b) Keep under review the content of this Protocol; (c) Consider and adopt any amendments to this Protocol or its annexes; (d) Consider reports prepared by the Secretariat on matters relating to the implementation of this Protocol; (e) Seek, where appropriate, the technical and financial services of relevant international bodies and scientific institutions for the purpose of the objective of this Protocol; (f) Establish such subsidiary bodies as may be deemed necessary for the implementation of this Protocol; (g) Fulfil the tasks as described in Article 4, paragraph 3, of this Protocol; (h) Consider strategies, action plans and programs for the implementation of this Protocol; (i) Perform such other functions, as may be required for the implementation of this Protocol.* Also, the EIA Protocol (Article 13), Ashgabat Protocol (Art. 12, Para. 1), and Moscow Protocol (Article 19, Para. 1) provide institutional obligations for their implementation assigned to the Conference of Parties established in the framework of the Tehran Convention.

---

[74] CITES, Article XIII.

[75] Attachment IV Montreal Protocol 1987.

10.6.2 Secretariat

The Secretariat is a collective treaty body that consists of an executive secretary, who is appointed by the Conference of the Parties (Article 22, Section 10h) as the chief administrative officer of the Secretariat (Article 23, Section 3), as well as necessary staff. The Secretariat shall provide formalized and institutionalized monitoring by determining whether the parties carry out their contractual obligations. Its supervisory functions are reflected in Article 23 of the Tehran Convention. The Secretariat has to create certain compliance incentives and ensure respective assistance. It shall prepare and transmit to the contracting parties notifications, reports, and other information received. This task should not be limited to the mere receipt and forwarding of the reports but shall include the processing of reports for other organs. The Secretariat shall prepare and transmit reports on matters relating to the implementation of this Convention and its protocols, as well as consult states on matters relating to the implementation of this Convention and its protocols.

The Secretariat shall assist states in complying with the Convention and its Protocols. It is aimed at tackling the root causes of poor or nonfulfillment of specified obligations, which often originates from the vagueness of material obligations in the Convention or shortcomings in the infrastructure for the performance in line with the Convention. Compliance assistance, explicitly mentioned by the Tehran Convention, includes considering enquiries and information from the contracting parties and consulting them on matters relating to implementation (Article 23, Para. 4c), as well as arranging, upon request from any contracting party, for the provision of technical assistance and advice for effective implementation (Article 23, Para. 4f). Assistance with capacity building for environmental purposes, providing among other things legislative support, is nowadays a common practice.[76] Also, the transfer of technology, understood as supporting scientific and technological development, is related to capacity building and is globally recognized as necessary.[77]

The function of the Secretariat established under the Tehran Convention is provisionally executed by UNEP, which is based in Geneva. This task was overtaken by UNEP in July 2004 in accordance with a request made by parties to the Tehran Conference. During the Fifth Caspian Summit on August 12, 2018, it was instructed to finalize the process of locating the Secretariat of the Convention in the territory of the Caspian states on a rotation basis.[78]

The additional protocols to the Tehran Convention foresee institutional obligations for their implementation, assigned to the Secretariat, established in the framework of the Tehran Convention (Article 13, Para. 2, of the EIA Protocol; Article

---

[76] Article 4 (2) Vienna Convention for the Protection of the Ozone Layer; Chapter 33, 34 Agenda 21.
[77] Article 202, 203, 266, 267 UNCLOS; Article 4, 5 Climate Change Convention.
[78] TC/cop6/info5 Framework Convention for the Protection of the Marine Environment of the Caspian Sea, 6–8 November 2019, Baku, Azerbaijan.

12, Para. 2, of the Ashgabat Protocol; and Article 19, Para. 2 of the Moscow Protocol).

10.6.3 Specific Institutional Arrangements

Apart from the Conference of Parties and the Secretariat as the main institutions responsible for the implementation of the Tehran Convention and its additional protocols, there is one more specific institutional arrangement envisaged in the context of regional preparedness, response, and cooperation in combating oil pollution accidents. According to the Aktau Protocol from 2011, which is the only additional protocol to the Tehran Convention already in force, there is a special regional mechanism envisaged (Article 13). The goals of a regional mechanism are to assist states in the prompt and effective reaction to oil pollution incidence. Regional mechanism includes the following functions (Article 13, Para. 2): *(a) Establishing close working relationships with the Competent National Authority of the Contracting Parties and also, where necessary, with relevant international and regional governmental and non-governmental organizations and bodies dealing with oil pollution incidents; (b) Coordinating regional activities with regard to technical co-operation, training, exercises, and providing expertise in cases of emergency, and assisting national activities in these fields; (c) Collecting and disseminating information on oil pollution incidents (inventories, expert opinions, reports on incidents, technical progress for improving contingency plans, etc.); (d) Preparing systematic procedures for data and information exchange concerning oil pollution incidents; (e) Acting as the focal point for exchanges of information on techniques for surveillance of oil pollution incidents in the Caspian Sea; (f) Making proposals on updating of the Caspian Sea Plan concerning Regional Co-operation in Combating Oil Pollution in Cases of Emergency; (g) Performing such other functions as may be required by the Conference of Parties.*

10.7 Procedures

The Tehran Convention sets it as the contracting parties' duty to "co-operate on a multilateral and bilateral basis in the development of Protocols to the Tehran Convention" (Article 6), which reflects the basic principle of environmental law. The practical importance of this principle has been already emphasized in a number of general political commitments[79] and binding international instruments,[80] as well

---

[79]Stockholm Declaration 1972, Principle 24; Rio Declaration 1992, Principle 27.
[80]Industrial Accident Convention 1992, Preamble.

as agreements of regional[81] and global application.[82] The general obligation of cooperation among the Caspian states is related to the matter of implementation of the Tehran Convention, providing that the protocols developed in a cooperative way should prescribe "additional measures, procedures and standards for the implementation" of the Tehran Convention (Article 6). However, more specific commitments have been provided for environmental impact assessment, rules concerning information exchange, consultation, and notification. Research programs require specific, not general, issues of concern.[83]

The same requirement for state cooperation—directly through the Secretariat of the Teheran Convention and with international organizations—regarding land-based sources of pollution is expressed in the Moscow Protocol (Article 16). Scientific and technical cooperation is important and should be encouraged by states. They should endeavor to cooperate, upon request for assistance, in developing scientific, technical, educational, and public awareness programs; training scientific, technical, and administrative personnel; as well as providing technical advice, information, and other assistance.

10.7.1 Exchange of Information

The availability of, and access to, environmental information is a condition for a successful and sustainable use, protection, and restoration and for cooperation within states as well as among them. A general obligation to exchange information is present in every international environmental treaty. With time, environmental information has become, step by step, the central issue of international environmental law.[84] There are already a lot of noteworthy international instruments regarding the issue of information.[85] To improve the flow of information and compliance with the

---

[81] London Convention 1933, Article 12(2); Western Hemisphere Convention 1940, Article VI; Alpine Convention 1991, Article 2(1).

[82] Vienna Convention 1985, Article 2(2), Biodiversity Convention 1992, Article 5.

[83] Article 20 research programs should be aimed, inter alia, at: (a) developing methods for the assessment of the toxicity of harmful substances and investigations of its affecting process on the environment of the Caspian Sea; (b) developing and applying environmentally sound or safe technologies; (c) the phasing out and/or substitution of substances likely to cause pollution; (d) developing environmentally sound or safe methods for the disposal of hazardous substances; (e) developing environmentally sound or safe techniques for water—construction works and water—regulation; (f) assessing the physical and financial damage resulting from pollution; (g) improvement of knowledge about the hydrological regime and ecosystem dynamics of the Caspian Sea including sea level fluctuations and the effects of such fluctuations on the Sea and coastal ecosystems; (h) studying the levels of radiation and radioactivity in the Caspian Sea.

[84] Stockholm Declaration 1972, Principle 2; 1982 World Charter for Nature, Paras. 15, 18, 19, 23.

[85] IAEA Notification Convention 1986; Basel Convention 1989; EC Directive on Environmental Information 1991; Industrial Accidents Convention 1992, Rio Declaration 1992; Agenda 21, Chapter 40.

more general objective to exchange information, a number of conventions have developed more strict procedures, such as the establishment of a documentation service,[86] an information service,[87] or a permanent committee of information.[88] The right of access to information on the environment, whether for the large public or particular groups, is a recent achievement of international environmental law, which includes a citizen's right of access to information.

The principle of accessibility of information on the pollution of the marine environment of the Caspian Sea is mentioned as one of the basic and fundamental rules of the Tehran Convention. Accordingly, the contracting parties provide each other with as much relevant information as possible:

> The principle of accessibility of information on the pollution of the marine environment of the Caspian Sea according to which the Contracting Parties provide each other with relevant information in the maximum possible amount (Article 5, Para. c)

The determination of the Caspian littoral states to comply with the obligation to inform and grant access to information about the pollution of the marine environment of the Caspian Sea reflects a significant progress compared to the often reserved attitude of the Caspian littoral states regarding the internationally recognized obligation of providing information on environmental damage. A requirement for states to continuously collect and exchange environmental data is provided for in the Tehran Convention:

> The Contracting Parties shall directly or through the Secretariat exchange information, on regular basis, in accordance with the provisions of this Convention (Article 21, Para. 1).

This explicit term "exchange of information" heads Article 21 of the Tehran Convention. However, this article requires that the contracting parties exchange information "on a regular basis," which in international environmental law used to be interpreted as an obligation to report—to provide regular or periodic information on specified matters to a specified body. That is why "reporting" may be distinguished from the usual meaning of the term "information exchange."

General information for the special purposes of another state should be provided, especially technical and scientific information. This legal obligation shapes Article 20 of the Tehran Convention, requiring cooperation among the contracting parties "in the conduct of research and development of effective techniques" concerning all issues of pollution of the Caspian Sea and "to endeavor to initiate or intensify specific research programs." The required cooperation includes the obligation to provide general information on one or more matters to another state on an ad hoc basis, especially regarding issues of research and development, even if this article does not explicitly mention it.

The Moscow Protocol required Caspian states (Article 14) to exchange available data and information on the state of the marine environment and coastal areas

---

[86] European Plant Protection Convention 1951, Article V(9).
[87] South-West Asia Locust Agreement 1963, Article II(1).
[88] African Phyto-Sanitary Convention 1954, Article 9.

regarding pollution from land-based sources on a regular basis, directly or through the Secretariat. The contracting parties should also develop systems and networks for the exchange of information. Also, the EIA Protocol to the Tehran Convention requires states parties, to an extent possible, to promote the exchange of information (Article 23, Para. 4). The Ashgabat Protocol requires contracting parties to "adopt appropriate measures to facilitate the exchange of information from all publicly available sources relevant to the conservation on biological diversity and sustainable and rational use of biological resources. Such exchange of information shall include the results of technical, scientific and socio-economic research, as well as information on training and traditional knowledge" (Article 17, (a)).

In 2014, the Caspian states adopted the Agreement on Hydrometeorological Cooperation in the Caspian Sea, taking into account the special hydrometeorological conditions on the Caspian Sea due to periodic fluctuations in its level, as well as the intensification of economic activity in its basin (Preamble). The goal of the Agreement is the creation and development of an integrated regional system for receiving and exchanging information on the Caspian Sea conditions (Article 2). To achieve the goals of this Agreement, the parties shall:

(1) Respect for the sovereignty, territorial integrity, independence and sovereign equality of States, non-use of force or the threat of force, mutual respect, cooperation and non-interference into the internal affairs of each other; (2) Using the Caspian Sea for peaceful purposes, making it a zone of peace, good-neighbourliness, friendship and cooperation, and solving all issues related to the Caspian Sea through peaceful means; (3) Ensuring security and stability in the Caspian Sea region; (4) Ensuring a stable balance of armaments of the Parties in the Caspian Sea, developing military capabilities within the limits of reasonable sufficiency with due regard to the interests of all the Parties and without prejudice to the security of each other; (5) Compliance with the agreed confidence-building measures in the military field in the spirit of predictability and transparency in line with general efforts to strengthen regional security and stability, including in accordance with international treaties concluded among all the Parties; (6) Non-presence in the Caspian Sea of armed forces not belonging to the Parties; (7) Non-provision by a Party of its territory to other States to commit aggression and undertake other military actions against any Party; (8) Freedom of navigation outside the Territorial Waters of each Party subject to the respect for sovereign and exclusive rights of the coastal States and to the compliance with relevant rules established by them with regard to the activities specified by the Parties; (9) Ensuring safety of navigation (10) The right to free access from the Caspian Sea to other seas and the Ocean, and back in accordance with the generally recognized principles and norms of international law and agreements between the relevant Parties, with due regard to legitimate interests of the transit Party, with a view to promoting international trade and economic development; (11) Navigation in, entry to and exit from the Caspian Sea exclusively by ships flying the flag of one of the Parties; (12) Application of agreed norms and rules related to the reproduction and regulation of the use of shared aquatic biological resources; (13) Liability of the polluting Party for damage caused to the ecological system of the Caspian Sea; (14) Protection of the environment of the Caspian Sea, conservation, restoration and rational use of its biological resources; (15) Facilitation of scientific research in the area of ecology and conservation and use of biological resources of the Caspian Sea; (16) Freedom of overflight by civil aircraft in accordance with the rules of the International Civil Aviation Organization; (17) Conducting marine scientific research outside the Territorial Waters of each Party in accordance with legal norms agreed upon by the Parties, subject to the respect for sovereign and exclusive rights of the coastal States and to the compliance with relevant rules established by them with regard to certain types of research (Article 3).

Further negotiations on information exchange are being continued under the preparatory process for the development of a Protocol on monitoring assessment and information exchange in the Framework Convention for the protection of the maritime environment of the Caspian Sea. This Protocol will be presented at the end of Sect. 10.7.2 on monitoring.

10.7.2 Monitoring

The enforcement authorities use the following as a source of information on the implementation of environmental legislation: information provided by other contracting parties, data from nongovernmental organizations, onsite investigations, and finally monitoring.[89]

Monitoring entails a frequent obligation of the parties to international environmental agreements to collect information relevant to specific or general environmental commitments.[90] The monitoring serves a number of purposes, of which the most common is to support research and investigate trends that reflect the state of the environment.[91] The monitoring serves the procurement of data of a technical and scientific nature. These data support future legislation rather than present the implementation that has already taken place. Accordingly, appropriate databases are created that prepare, store, and make available to the public the collected information on the prevention of cross-border environmental damage.

The Tehran Convention also calls for a repeated measurement of the quality of the environment on the Caspian Sea, providing for a regular individual or joint assessment of the environmental conditions of the Caspian Sea and the effectiveness of measures taken for the prevention, control, and reduction of the pollution of its marine environment. To achieve this aim, parties should make appropriate effort to "harmonise rules for the setting up and operation of monitoring programmes, measurement systems, analytical techniques, data processing and evaluation procedures for data quality" (Articles 19.3 and 19.4). The obligation to "harmonise rules" does not include monitoring to ensure compliance with the objectives of the agreement, to avoid the involvement of another state in the compliance process.

The Tehran Convention does not directly provide for the involvement of international organizations in information gathering, which is now a widespread

---

[89] See: Beyerlin, p. 244.

[90] Antarctic Convention 1959, Article VII; London Convention 1972, Article VI(1)(d); Rhine Chemical Pollution Convention 1976, Article 10(1); Paris LBS Convention 1974, Article 11; 1976 Barcelona Convention, Article 101982 UNCLOS, Article 204(1) and (2) and Article 226 (1); OSPAR Convention 1992, Article 6 and Annex IV; Helsinki Convention 1992, Article 11; Biodiversity Convention 1992, Article 7(b) and (c); Agenda 21, Chapter 40.

[91] OSPAR Convention 1992, Annex IV, Article 1.

international practice.[92] However, the possibility to bridge this gap may be limited again by a party's obligation to act in accordance with standards commonly used in international practice regarding pollution issues.

The Tehran Convention provides that the contracting parties should develop a centralized database and information management system, which would serve as a repository for all important data and information, for decision-making, education, administration, and general public knowledge (Article 19.5). Therewith, this provision will also fulfill another international obligation requiring states to improve public education and awareness on environmental matters.[93] The Secretariat of the Convention is committed to establishing and maintaining the database of national and international laws relevant to the protection of the Caspian Sea (Article 23.4 (e)). The obligation to collect, compile, and evaluate in order to identify sources that are likely to cause pollution to the Caspian Sea is in the context of the general obligation to cooperate in the prevention, reduction, and control of pollution and the protection, preservation, and restoration of the marine environment of the Caspian Sea (Article 18.3(a)):

> The Contracting Parties shall co-operate in the formulation of an Action Plan for the Protection of the marine environment of the Caspian Sea in order to prevent, reduce and control pollution and to protect, preserve and restore the marine environment of the Caspian Sea (Article 18(2)).

The "National Caspian Action Plan" (NCAP) is elaborated by the Caspian littoral states, considering the assessment of the most vulnerable environmental areas that match with priority areas identified in the Transboundary Diagnostic Analysis for the Caspian Sea (TDA) of 2002. The purpose of conducting a TDA is to scale the relative importance of sources and causes, both immediate and root, of transboundary water problems and to identify potential preventive and remedial actions. In the NCAPs, each state developed goals and presented certain actions, as well as made available resources and strategies to achieve the planned objectives. The NCAPs were prepared before May 2002 and together with the TDA became the basis of the so-called Strategic Action Program (SAP) for the Caspian Sea, which was prepared in November 2003.[94] The SAP sets the agenda for enhanced regional environmental cooperation among the littoral states over the next 10 years, approximately 2007–2017, in two distinct 5-year periods. To improve environmental stewardship and protect the ecosystems of the Caspian, the SAP outlines five regional environmental quality objectives (EQOs) to be addressed and identifies environmental interventions taken to meet those EQOs at the national and regional levels. The SAP builds upon and complements the NCAPs. The SAP is the result of the regional consultation process involving both the Caspian littoral states and

---

[92]Earthwatch, Global Environmental Monitoring System (GEMS), International Environmental Information System (INFOTERRA), The European Environment Agency.
[93]Rio Declaration 1992, Principle 10; Agenda 21, Chapter 36.
[94]Strategic Action Program for the Caspian Sea, prepared by Caspian Environmental Programme, in: Programme Coordination Unit, 2003.

international partner organizations. In 2006, the need to update the TDA, SAP, and NCAP was acknowledged, which was based on the consistency of the four priority regional concern areas identified in the first SAP of 2003: first, unsustainable use of bioresources; second, threats to biodiversity, including those from invasive species; third, marine and coastal pollution; and, fourth, unsustainable coastal area development. The SAP defines the financial and institutional structures required for the implementation of the priority actions approximately till 2017.[95]

According to the Moscow Protocol (Article 13), the contracting parties, to an extent possible, should collect data and information and prepare and maintain a national database on the conditions of the marine environment and coastal areas of the Caspian Sea and on inputs of substances listed in Annex I of this Protocol from land-based sources. They should also undertake a regional assessment on a regular basis (at least once in 5 years) of the state of the marine environment and coastal areas of the Caspian Sea and collaborate in establishing elements of the regional monitoring program as well as compatible national monitoring programs. The EIA Protocol foresees a need for a postproject analysis. It shall be undertaken to achieve among others the objectives of monitoring and compliance set out in the authorization or approval offered by the authority, as well as to achieve effectiveness of mitigations measures (Article. 11, Para. 2(a)). Annex III contains a minimum required content of the environmental impact assessment documentation. It foresees that information included to the EIA shall include information reflected in Article 6, regarding monitoring and management programs. The Moscow Protocol to the Tehran Convention provides that *the Contracting Parties shall collaborate in establishing elements of the regional monitoring programme as well as compatible national monitoring programmes, with analytical quality control, and to promote data storage, retrieval and exchange* (Article 13, Para. 2).

Monitoring as well as assessment and information exchange are being negotiated among the Caspian riparian countries in the framework of the Draft Protocol on Monitoring Assessment and Information Exchange to the Framework Convention for the Protection of the Marine Environment of the Caspian Sea. The idea of this Protocol was for the first time discussed in 2014 at the fifth meeting of the Conference of Parties to the Tehran Convention. A draft of this Protocol is currently under review and should be finalized and submitted to the Conference of Parties' sixth meeting for adoption. It discussed the location of the monitoring stations, the type of medial-like water sediment and biota, minimum data quality, formats for data submission, information on certified laboratories monitoring activities, submission of data, and the organizations responsible for the accumulation, checking, compiling, and managing of data, as well as the definition of rights of data owners.

---

[95]See (Golubev 1997).

10.7.3 Environmental Impact Assessment

A project that could potentially cause significant transboundary environmental pollution and adversely affect the environment of another state is to be subjected to an environmental impact assessment. The function of EIA is to provide the concerned parties with information on the environmental effects of their decisions, to require decisions to be in accord with its provisions, and to ensure the participation of all potentially interested actors in the decision-making process. Since the 1970s, environmental impact has been progressively adopted into global and many national legal systems as a response by the international community to the request to devise strategies to hold and limit the effects of environmental degradation.[96] There are some binding international acts explicitly requiring the application of the environmental impact assessment, which were preceded by other agreements providing for the explicit or implied general obligations on environmental impact assessment.[97] The obligation of states to warn each other in a timely manner in case of a transboundary environmental accident is gradually becoming part of customary law.[98] The countries of Europe and America have created their own criteria of transboundary environmental impact assessment in their national legislation. These criteria, however, are often not only limited in scope but also do not cover all transboundary actions that could potentially have an adverse effect, and neither do they provide for detailed procedural mechanisms of the EIA. The EIA is to contribute to an effective environmental protection through the identification, description, and assessment of the impact of policies on the environment. Within an IEA procedure, various methods of analysis, forecasting, evaluation, participation, and cooperation can be used. The test should be carried out systematically according to a

---

[96] 1972 National Environmental Protection Act; Rio Declaration 1992, Principle 16; Stockholm Declaration 1972, Principle 14 and 15; OECD Council Recommendation C(74) 216, Analysis of the Environmental Consequences of Significant Public and Private Projects, 14 Nov. 1974; OECD Council Recommendation C(79)116, Assessment of Projects with Significant Impact on the Environment, 8 May 1979; FAO Comparative Legal Strategy on Environmental Impact Assessment and Agricultural Development, 1982 FAO Environmental Paper; OECD Council Recommendation C(85) 104 on Environmental Assessment of Development Assistance Projects and programs, 20 June 1985; 1986 World Commission on Environment and Development, Environmental Protection and Sustainable Development: Legal Principles and Recommendations, 58 to 62; Goals and Principles of Environmental Impact Assessment, UNEP/GC/DEC/14/25 (1987); 1992 Agenda 21, Paras. 7.41(b); 8.4; 8.5(b); 10.8(b); 9.12(b); 11.24(a); 13.17(a); 15.5(k), 16.45(c); 17.5 (d); 18.22(c); 19.21(d); 21.31(a); 22.4(d); 23.2.

[97] 1985 adopted EC Directive on Environmental Impact Assessment; ESPOO Convention; Antarctic-Environmental Protocol 1991 to the Antarctic Treaty of 1959.

[98] See: Pulp Mills on the River Uruguay (Arg. v. Uru.), 178–180 (Judgment of April 20, 2010), available at http://www.icj-cij.org/docket/files/135/15877.pdf. Accessed 1 July 2014. The Court observed that the practice of environmental impact assessment (EIA) "has gained so much acceptance among States that it may now be considered a requirement under general international law to undertake an environmental impact assessment where there is a risk that the proposed industrial activity may have a significant adverse impact in a transboundary context, in particular, on a shared resource," 204.

certain pattern. The effects of an intervention are to be evaluated gradually according to each analytical criterion.

For the success of the IEA, the full involvement of the public is of indisputable importance. Complex consultations and public debates improve the decisions taken as they provide several alternatives. Despite that, involving the public in national and regional legislation is going slowly.[99]

In the "Almaty Declaration on Cooperation in the field of Environmental Protection of the Caspian Sea Region" (1994), the Caspian coastal states decided to take coordinated measures to prevent the growing harmful transboundary impact on the Caspian marine environment.[100] International organizations were involved in these actions. For example, the objective of preventing risks associated with the exploitation of the energy resources of the Caspian Sea can be achieved with the help of EIA. The legal model for this regulation of the Tehran Convention was the Espoo Convention, which might be used as a standard for the development of regional EIA provisions. A detailed elaboration of the coordination measures for information exchange and cooperation of states in this respect would show a number of benchmark for the case of the Caspian Sea.[101] The Espoo Convention finds direct application merely to Azerbaijan and Kazakhstan, if considering all Caspian littoral states.[102]

EIA was recognized also by the Tehran Convention. According to the general duty of notification, the states should develop common standards that can ensure that transboundary environmental impact is prevented as far as possible. Without this procedural safeguard, the set objectives for the prevention, reduction, and control of pollution of the marine environment, as well as the protection, prevention, and restoration of the marine environment of the Caspian Sea, would stay ineffective (Article 4 (a), (b)). The Convention provides that "each contracting party shall take all appropriate measures to introduce and apply procedures of environmental impact assessment of any planned activity, that are likely to cause significant adverse effect on the marine environment of the Caspian Sea" (Article 17(1)). "The contracting parties are required to disseminate the results of environmental impact assessment and co-operate in the development of Protocols that determine procedures regarding this issue" (Article 17(2)).

For taking preventive and combating measures against pollution incidents, the Tehran Convention calls upon the contracting party of origin to identify hazardous activities within its jurisdiction that can potentially cause environmental emergencies. This party should ensure that other contracting parties are notified of any such proposed or existing activities. The contracting parties agreed to implement

---

[99]See: (Tilleman 1995).

[100]UNDP; WB; UNEP.

[101]Mediterranean Emergency Protocol 1976, Article 1; Kuwait Emergency Protocol 1978, Article II; Jeddah Pollution Emergency Protocol 1982, Article II; etc.

[102]Azerbaijan (25.03.1999 accession), Kazakhstan (11.01.2001 accession), Russia (6.06.1991 signature).

risk-reducing measures (Article 13, Section 2). This provision refers in part to the regulation of the UN Convention on the Law of Non-navigational Uses of International Watercourses of 1997. Its Article 12 states that before a watercourse state implements or permits the implementation of planned measures that may have a significant adverse effect upon other watercourse states, it should provide those states with a timely notification thereof. In contrast, the Tehran Convention speaks merely of already-existing cross-border harmful actions.

In the Tehran Convention, the ancillary Protocol on Environment Impact Assessment in a Transboundary Context (further referred to as EIA Protocol) was adopted on July 12, 2018, and signed by all contracting parties in Moscow. It defines environmental impact assessment (EIA) as a national procedure for evaluating the likely impact of a proposed activity on the environment (Article 1):

> "Proposed activity" means any activity or any major change to an activity subject to a decision of a competent authority in accordance with an applicable national environmental impact assessment procedure; "Impact" means any effect caused by implementation of a proposed activity on the marine environment of the Caspian Sea including flora, fauna, soil, atmospheric air, water, climate, landscape, historical monuments and/or interaction among those factors; also includes effects on human health and safety, cultural heritage, socio-economic or other conditions resulting from alterations to those factors;

> The objective of this Protocol is to implement effective and transparent environmental impact assessment procedures in a transboundary context to any proposed activity which is likely to cause significant transboundary impact on the marine environment and land affected by proximity to the sea in order to prevent, reduce and control pollution of the marine environment and land affected by proximity to the sea, promote conservation of its biodiversity, and rational use of its natural resources, and protect human health (Article 2).

The Protocol contains a list in Annex 1 of activities that are likely to cause significant transboundary impact, which shall be notified as early as possible by the state of origin of this activity to the state that is likely to be affected by the transboundary impact of the proposed activity. The Protocol describes in Article 5 the process of notification and communication between concerned parties. It follows with provisions on the preparation (Article 7) and review (Article 8) of the environmental impact assessment documentation and the public consultation transmittal of the draft EIA documentation. It also regulates consultations between concerned parties (Article 9) and the manner of adopting the final decision on the implementation of the proposed activity (Article 10) taken by the competent authority of the party of origin of the EIA after taking into account the comments received in the process of review of EIA documentation and public consultation (Article 10). Annex I to this Protocol lists activities that are likely to cause significant transboundary impact and that shall be subject to EIA procedure (Article 4), including:

> 1) Crude oil refineries (excluding undertakings manufacturing only lubricants from crude oil) and installations for the gasification and liquefaction of 500 tons or more of coal or bituminous shale per day. 2) Thermal power stations and other combustion installations with

a heat output of 300 megawatts or more. 3) Nuclear power stations and other nuclear reactors, including the dismantling or decommissioning of such power stations or reactors (1/) except research installations for the production and conversion of fissionable and fertile materials, whose maximum power does not exceed 1 kilowatt continuous thermal load. 4) Installations solely designed for the production or enrichment of nuclear fuels, for the reprocessing or storage of irradiated nuclear fuels or for the storage, disposal and processing of radioactive waste. 5) Major installations for the initial smelting of cast iron and steel and for the production of non-ferrous metals. 6) Installations for the extraction of asbestos and for the processing and transformation of asbestos and products containing asbestos: for asbestos-cement products, with an annual production of more than 20,000 tons finished product; for friction material, with an annual production of more than 50 tons finished product; and for other asbestos utilization of more than 200 tons per year. 7) Integrated chemical and petrochemical installations. 8) Construction, reconstruction and/or widening of motorways, express roads (2/) and lines for long distance railway traffic, including the construction of major associated bridges, and of airports (3/) with a basic runway length of 2,100 metres or more; 9) Large diameter pipelines for the transport of oil, gas and oil products, or chemicals. 10) Marine/Trading ports and also inland waterways and ports for inland-waterway traffic which permit the passage of vessels of over 1,350 tons. 11) Waste-disposal installations for the incineration, chemical treatment or landfill of waste; 12) Large dams, reservoirs and canals connected to the Caspian Sea. 13) Groundwater abstraction activities or artificial groundwater recharge schemes where the annual volume of water to be abstracted or recharged amounts to 10,000,000 cubic metres or more. 14) Pulp, paper and board manufacturing of 200 air-dried tons or more per day. 15) Major quarries, mining, on-site extraction and processing of metal ores or coal. 16) Offshore hydrocarbon production. Extraction of petroleum and natural gas for where the amount extracted exceeds 500 tons/day in the case of petroleum and 500,000 cubic metres/day in the case of gas. 17) Major storage facilities for petroleum, petrochemical and chemical products. 18) Deforestation of large areas. 19) Works for the transfer of water resources between and within river basins where this transfer aims at preventing possible shortages of water and where the amount of water transferred exceeds 100,000,000 cubic metres/year; in all other cases, works for the transfer of water resources between and within river basins where the multi-annual average flow of the basin of abstraction exceeds 2,000,000,000 cubic metres/year and where the amount of water transferred exceeds 5 per cent of this flow. In both cases transfers of piped drinking water are excluded. 20) Waste-water treatment plants with a capacity exceeding 150,000 population equivalent. 21) Installations of microbiological and biotechnological production, and the release of genetically modified organisms. 22) Land reclamation, including the construction of artificial islands, spits and reefs.

Provisions related to transboundary impact were included also in the Moscow Protocol to the Tehran Convention (Articles 11–12). In case of pollution from land-based sources and activities originating from the territory of the contracting party likely to have an adverse effect on other contracting states, the Protocol required countries to share information and enter into consultation. Additionally, states are required to adopt regional and national guidance concerning the assessment of the potential environmental impact of land-based activities on its territory. The application of EIA procedures to planned land-based activities likely to cause an adverse effect to the Caspian environment should ensure that the implementation of such activities will be conducted after having fully considered the outcomes of the EIA procedure and with prior written authorization from the country. The Moscow Protocol provides that contracting parties shall *promote cooperation between and among Contracting Parties in environmental impact assessment related to activities which are likely to have significant adverse effect on the marine environment of the Caspian Sea* (Article 4, Para. 2 (c)). The same Protocol envisages the need for environmental impact assessment (Article 12): *2. Each Contracting Party shall introduce and apply procedures of environmental impact assessment of any planned land-based activity or project within its territory that is likely to cause significant adverse effect on the marine environment and coastal areas of the Caspian Sea. 3. The implementation of activities and projects referred to in paragraph 2 of this Article shall be made subject to a prior written authorisation from the competent authorities of the Contracting Party, which takes fully into account the findings and recommendations of the environmental impact assessment.*

Also, the Ashgabat Protocol requires countries to *apply the procedures of environmental impact assessment as a tool of preventing and minimizing adverse impact* (Article 13).

10.7.4 Reporting

The objective of the states' obligation to report is to facilitate the implementation of an act in a particular contracting state. The Tehran Convention on the Protection of the Marine Environment of the Caspian Sea, in line with numerous international environmental agreements, provides for the reporting obligation of all parties regarding the implementation of the Convention and its protocols and regulates the scope of actions agreed upon by the Conference of the Parties:

> Each National Authority shall submit to the Secretariat reports on measures adopted for the implementation of the provisions of this Convention and its Protocols in format and at intervals to be determined by the Conference of the Parties. The Secretariat shall circulate the received reports to all Contracting Parties (Article 27).

According to international law, reporting is aimed at the implementation of a legal act. It differs significantly from the obligation to exchange information, which is aimed at exchanging scientific and technical data. The reporting obligation set in the Tehran Convention is of a more general nature, and it does not call upon states to provide detailed data. It requires neither ensuring of a comprehensive review of national legislation or administrative regulations, procedures, or practices or strategies for the implementation of legal acts nor the monitoring of the actual situation with regard to the implementation obligation nor the facilitation of exchange of information between the states parties with regard to achieving the objectives and rights and obligations set out in the Convention. There is no single model for how the reports specified in the Tehran Convention should be prepared that would ensure the consistency and comparability of the data. Data management was entrusted to competent national authorities. However, they often come across the problem of not possessing the required data and needing to request them from national information systems. The Treaty Conference is mandated to receive reports and to evaluate and determine the state of the environment on the basis of these reports.

There are a few types of reports among international environmental agreements that are also found in the Tehran Convention. First of all, some environmental agreements require institutional organs to provide reports to the parties informing them about activities taking place under the auspices of the treaty.[103] The Tehran Convention requires the Secretariat of the Convention to prepare and transmit to the contracting parties reports and other information received (Article 23.4(b)), as well as to "circulate the received report to all Parties" (Article 27). The format and intervals of reports will be decided by the Conference of the Parties. However, there is no indication of a particular time for the submission of the reports or how often they should be submitted. Therefore, there is the legal text requiring reports to relate to the provisions of this Convention and its protocols (Article 23.4(d)).

Another reporting obligation to be fulfilled is providing a report to the institutions established under the treaty. In the Tehran Convention, there are clear provisions regarding this issue. Each national authority is required to submit to the Secretariat reports on measures adopted for the implementation of the provisions of the Tehran Convention and its protocols (Article 27). The Conference of the Parties receives reports submitted by the contracting parties (Article 22.10(d)). Inter alia, the Conference is due to review and evaluate the state of the maritime environment and, in particular, the state of pollution and its effects, which can be used to evaluate the information the parties are obligated to provide.

To determine further details on the mandatory content of the reports, one must refer to the basic duties of the contracting parties to submit reports to the national authorities, as well as their obligation of cooperation to prevent, reduce, and control the pollution of the Caspian Sea, and by considering the requirements commonly used in international practice and the provisions of other international environmental treaties, we can extract other details regarding the content of the parties' reports. The reports require detailed and regular information and are used to provide information on the implementation of treaty commitments. Parties are required to provide reports on the establishment of any natural reserves,[104] authorization to issue licenses,[105] implementation measures and their effectiveness,[106] other environment and development issues they find relevant,[107] new and additional finance resources, access to environmentally sound technologies and know-how, etc.[108]

The other type of reporting is a report on environmental emergencies, which is required from nongovernmental actors.[109] A similar commitment is included in the Tehran Convention, which insists that the Conference of the Parties should state the

---

[103]Tropical Tuna Commission Convention 1949, Article 1(2); African Phyto-Sanitary Convention 1954; Article 3(b); Agreement establishing the EBRD 1990, Article 35.

[104]London Convention 1933, Article 5(1) and 8(6).

[105]International Whaling Convention 1946, Article VIII(1).

[106]Plant Protection Agreement 1956, Article II(1)(b); Basel Convention 1989, Article 3(1); Biodiversity Convention 1992, Article 26; Climate Change Convention 1992, Article 12; OSPAR Convention 1992, Article 22.

[107]UNGA Res. 47/191 (1992), Para. 3(b).

[108]Climate Change Convention 1992, Articles. 12(3), and 4(3), (4) and (5).

[109]Agenda 21 UNGA res 47/191 (1992), Para. 3(h).

extent of pollution and its effects based on reports provided by both the contracting parties and any competent international or regional organization (Article 22.10 (d)).

Based on the parties' obligation to act in accordance with standards commonly used in international practice by formulating, elaborating, and harmonizing rules to prevent, reduce, and control pollution, contracting parties are also required to provide reports of an event, other than an emergency situation, that may entail a significant environmental risk. The need for such an obligation has been widely recognized in international environmental law since the mid-1970s.[110]

According to the Moscow Protocol (Article 17), states are obliged to submit a report on the implementation of the protocol provisions regarding land-based pollution to the Secretariat of the Teheran Convention, which is to prepare respective regional reports. The EIA Protocol requires each contracting party to submit a report to the Secretariat on the implementation of the provisions of the Protocol (Article 12). In the Moscow Protocol, each contracting party is required to report to the Secretariat on the implementation measures (Article 17).

10.7.5 Consultations

The Tehran Convention for the Protection of the Marine Environment of the Caspian Sea does not refer expressly to the obligation of consultation. However, the Caspian littoral states are obliged to consult, according to the general international practice of states so that the pollution of the Caspian Sea remains below the threshold of significance.

The obligation to consult may occur for different reasons, which may all arise from issues regarding the Caspian Sea. Consultation may be required, first, on the implementation of a treaty;[111] second, when the proceeding of one state may cause considerable disruption to the environment or to the rights of another state;[112] third, concerning the use of shared natural resources;[113] and, fourth, in time of emergency.[114]

The obligation to consult is connected with the principle of "prior informed consent," which is currently adopted and widely recognized by a number of

---

[110]Stockholm Declaration 1972; UNEP draft Principle of Conduct 1978, Principle 6; Rio Declaration 1992, Principle 19; 1980 Agreement between Spain and Portugal on Co-operation in Matters Affecting the Safety of Nuclear Installation, Article 2.

[111]ASEAN Agreement 1985, Article 18(2)(e).

[112]Quito LBS Protocol 1983, Article XII and Athens LBS Protocol 1980, Article 12(1); 1974 Nordic Environmental protection Convention, Article 11 and Espoo Convention 1991, Article 5 and Industrial Accidents Convention 1992, Article 4; 1979 LRTAP Convention, Article 5.

[113]1968 African Nature Convention, Article V(2); Ramsar Convention 1971, Article 5; 1982 UNCLOS, 142(2); 1982 Geneva SPA Protocol, Article 6(1).

[114]1981 Abidian Emergency Protocol, Article 10 (1)(b); London Convention 1972, Article V(2); 1986 I.E. Notification Convention, Article 6; ILO Radiation Convention 1960, Article 1.

international legal instruments.[115] As a result of an accident like the Chernobyl disaster, a number of treaties,[116] nonbinding instruments,[117] and state practice[118] now include commitments of states to provide an emergency notification of incidents that are likely to have a significant effect on the environment. A consequence of the Chernobyl accident was the widely spread opinion that the obligation to notify in case of emergency situations is a rule of international law. It does not apply to military nuclear accidents. It has already been recognized by international courts and tribunals[119] and in a number of instruments of international environmental law.[120] Also, notification of an emergency situation is required. Some treaties not only provide for the requirement that states consult with each other and with nongovernmental actors but require that a consultative committee should be appointed.[121] In the Caspian Sea case, its implementation would be possible by a decision of the Conference of the Parties to "establish such subsidiary bodies as may be deemed necessary for the implementation of the Convention and its Protocols."[122]

In terms of consultations between concerned parties, the EIA Protocol provides in Article 9 the following:

> Prior to making the final decision on the proposed activity, at the request of the Affected Party, the Party of Origin shall enter into consultations with the Affected Party, concerning, inter alia, measures to reduce potential transboundary impact (Article 9, Para. 1).
>
> The Concerned Parties shall agree, at the commencement of such consultations, on a reasonable time-frame for the duration of the consultation period, while the period of consultations should not exceed 180 days, unless otherwise decided during the consultation period (Article 9, Para. 2).

The EIA Protocol to the Tehran Convention specifies that in case there is no notification from the party of origin and another contracting party has reasonable concerns about the significant transboundary impact to be felt from the activity proposed by the party of origin, both parties, if necessary, shall hold a consultation regarding a possible participation in an environmental impact assessment procedure (Article 5, Para. 9).

---

[115] 1985 FAO Pesticides Guidelines; 1989 UNEP London Guidelines; Basel Convention 1989; 1989 Lomé Convention.

[116] 1982 UNCLOS, Article 198; 199 Biodiversity Convention, Article 14(1)(d); 1992 Rio Declaration 1992, Principle 18.

[117] 1974 OECD Recommendation, Para. 9; 1978 UNEP draft Principles of Conduct, Principle 9.

[118] 1982 Montreal ILA Rules, Article 7; 1987 IDI Resolution, Article 9(1)(a).

[119] Lac Lanoux Arbitration, 24 I.L.R. 101 (1957); Fisheries Jurisdiction Cases (United Kingdom v. Iceland) (Merits), 1974 ICJ Rep. 3, Special Agreement between Hungary and the Slovak Republic for Submission to the ICJ of the Differences between them, 32 I.L.M. 1294 (1993).

[120] 1978 UNEP Draft Principles, Principle 7; 1986 WCED Legal Principles, Article 17; Rio Declaration 1992, Principle 19, etc.

[121] Treaty of Rarotonga 1985, Article 10 and Annex 3.

[122] Article 22.10(g) of Tehran Convention.

10.7.6 Public Access to Information

The norms of the Tehran Convention for the Protection of the Marine Environment of the Caspian Sea with regard to the obligation to provide information does not restrict the right to environmental information only to the states but are also aimed at informing the citizens and the general public:

> The Contracting Parties shall endeavour to ensure public access to environmental conditions of the Caspian Sea, measures taken or planned to be taken to prevent, control and reduce pollution of the Caspian Sea in accordance with their national legislation and taking into account provisions of existing international agreements concerning public access to environmental information (Article 21(2)).

In this regard, the Tehran Convention builds upon the international standards as a benchmark for the regional rules on information access. The citizens' right of access to environmental information can be found in numerous international agreements.121 European law extends this right and awards citizens the following rights: access to information, public participation in decision-making, and access to justice in environmental matters explicitly settled in the so-called "Aarhus Convention" of the United Nations Economic Commission for Europe (UNECE).122 According to the Aarhus Convention (Article 2, Section 3), "environmental information" means any information in written, visual, aural, electronic, or any other material form on the following: first, the state of environmental elements, such as air and atmosphere, water, soil, land, landscape and natural sites, and biological diversity and its components, including genetically modified organisms, and the interaction among these elements; second, factors such as substances, energy, noise and radiation, activities or measures, including administrative measures, environmental agreements, policies, legislation, plans, and programs, affecting or likely to affect the elements of the environment, and cost-benefit and other economic analyses and assumptions used in environmental decision-making; third, the state of human health and safety, the conditions of human life, cultural sites, and built structures, inasmuch as they are or may be affected by the state of the environmental elements or, through these elements, by the factors, activities, or measures referred to above.

Three Caspian littoral states, the former Soviet republics of Azerbaijan, Kazakhstan, and Turkmenistan, are parties to the Aarhus Convention and are obliged to apply all its provisions. However, the scope of the assurance of access to information, public participation in decision-making, and access to justice in environmental matters is more limited in its scope. The Tehran Convention emphasizes the parties' obligation to ensure public access to the environmental conditions of the Caspian Sea and to take appropriate measures to prevent, control, and reduce the pollution of the Caspian Sea (Article 21.2). These endeavors should take place in accordance with the party's national legislation and should consider provisions for existing international treaties regarding public access to environmental information. The Tehran Conventions requires parties merely to "endeavour to ensure public access to environmental conditions of the Caspian Sea." In contrast, in its Article 1, the Aarhus Convention states that "each party shall guarantee" all three forms of access to environmental information. Parties to the Tehran Convention merely agreed to "consider" the international standards of public participation, among others, the Aarhus Convention. In fulfilling this obligation, the Caspian littoral states are subordinated to their own national law and shall act in accordance with it. As for Azerbaijan, Kazakhstan, and Turkmenistan, which are parties to the Aarhus Convention, they are obliged to intensify the public's participation in environmental matters and to strengthen the environmental awareness of the population also with respect to the affairs of the Caspian Sea. The three pillars of the Aarhus Convention—access to information, public participation in decision-making, and access to justice in environmental matters—should find implementation in the case of the Caspian Sea.[123]

The implementation of and compliance with other international environmental rules such as eco-labeling (the labeling of the environmental aspects of goods and services),[124] eco-accounting (to assess if the environmental costs are properly accounted for by both individual firms and states and that information about them is easily accessible),[125] and auditing (the technique of allowing firms or states to conduct the eco-accounting).[126] The internationally recognized problem of the limited effectiveness of these obligations is partly caused by the unwillingness of states to share information of potential commercial value or that could violate intellectual property rights.[127] According to the MOSCOW Protocol (Article 15), it promotes the participation of local authorities and the public in measures that are necessary for the protection of the marine environment and coastal areas of the Caspian Sea against pollution from land-based sources and activities. States should also facilitate public access to information concerning the conditions of the marine environment and coastal areas of the Caspian Sea and the measures taken or planned to be taken to prevent, control, and reduce pollution by considering the provisions of existing international agreements concerning public access to environmental information.

Also, the concerned parties shall discuss and agree on the format and languages to be used for the environmental impact assessment documentation to be submitted for the purposes of public consultations (Article 6, Para. 2 (a)). Also, the EIA Protocol provides that the party of origin shall ensure that the project proponent prepares a draft of the environmental impact assessment documentation for the purposes of public consultations (Article 7). The EIA Protocol prescribes that concerned parties shall ensure that the public in the areas likely to be affected have the opportunity to participate in the public consultations (Article 8). Also, the Ashgabat Protocol states that contracting parties shall *endeavour to promote the participation of public and*

---

[123]See: (McAllister 1998).

[124]Council Regulation 92/880/EEC, OJ L 99, 11 April 1992, Article 1.

[125]Report of the Secretary General: Accounting for Environmental Protection measures; UN doc. E/C.10/1991/5, 11 February 1991.

[126]Council Regulation 93/1836/EEC, OJ L 168, 10 July 1993, 1; Article 1(1) and (2).

[127]Biodiversity Convention, Article 16.

conservation organizations in measures necessary for the protection of protected areas and threatened species. (Article 18 (c)).

10.8 Implementation of the Tehran Convention and Compliance

10.8.1 Compliance

In the Teheran Convention (Article 28), the Caspian countries agreed to cooperate in the development of procedures to ensure compliance with the provisions of this Convention or its protocols. Instruments requiring a more detailed compliance with the regulations on land-based sources of pollution include the Moscow Protocol (Article 18). It prescribes that the Conference of the Parties shall review and evaluate parties' compliance with the Protocol and the decisions and recommendations adopted thereunder based on the received reports and any other information submitted by the contracting parties and, where appropriate, decide and call for steps to bring about compliance with the Protocol and the decisions adopted thereunder and promote the implementation of recommendations, including measures to assist a contracting party to carry out its obligations.

10.8.2 Liability and Compensation

The rules regarding states' liability for environmental damage need to be considered in relation to international treaties, customary law, and general law principles. Customary law creates an obligation to avert any damage to the environment. But the first problem is defining environmental damage. There are no common international norms regarding this issue; however, some treaties make a certain contribution toward this aim, defining "pollution"[128] or "adverse effects"[129] as having of marginal value in causing liability for environmental damage. The International Court of Justice, in the *Trail Smelter* case, required a "serious consequence" to justify the claim.[130] Environmental damage does not include damage to persons or property.[131] However, the establishment of a threshold of tolerable environmental damage may vary according to local terms. Respectively, states set their own environmental

---

[128]UNCLOS 1982, Article 1(4).
[129]Vienna Convention 1985, Article 1(2); Climate Change Convention, Article 1(1).
[130]Trail Smelter Case, 16 APRIL 1938, 11 March 1941;3 R.I.A.A 1907 (1941).
[131]ILC Draft Articles on International Liability, Article 24.

standards. Some guidelines in this respect have been prepared, for instance, by the European Commission[132] and the World Health Organization (WHO).[133]

The issues of defining an international standard of care applicable to the obligation of preventing environmental damage, fault,[134] and strict liability[135] and also whether absolute liability exists[136] also remain unresolved.

The obligation to make amends for the consequences of an illegal act is well established.[137] The sanctions for the abuse of an international environmental obligation could be a formal apology,[138] a declaration by an international tribunal recognizing the legal position of the other party,[139] or punishment of the guilty persons. Another problem concerns the basis for assessing the extent of environmental damage or the cost of reinstatement, as well as a theoretical model for calculation. This aspect of international law is not developed; however, the *Trail Smelter* case and limited state praxis have established a precedent.

Some international treaties provide for state liability for environmental damage.[140] The International Law Commission (ILC) has prepared draft liability articles, trying to establish basic principles concerning this issue.[141] States' liability for environmental damage is considered not well developed and as requiring further building up. The reference made by the Tehran Convention to the principles and norms of international law on the issue of liability and compensation for environmental damage seems to be regarded as giving assistance in the long-term perspective but not immediately.

Under provisions relating to "liability and compensation for damage to the environment of the Caspian Sea resulting from violations of the provisions of this Convention and its Protocols," the Tehran Convention refers to principles and norms of international law (Article 28). It provides for the fact that, on this basis, contracting parties should undertake to develop appropriate rules and procedures. However, although the general principles of international law concerning states' liability are relatively well developed, a lot needs to be done with regard to the environmental damage caused by states.[142]

---

[132] Radiological Protection Criteria for Controlling Doses to the Public in the Event of Accidental Releases of Radioactive Material, A Guide on Emergency Reference Levels of Dose from the Group of Experts Convened under Article 41 of the EURATOM Treaty (1982).

[133] Nuclear Power: Principles of Public Health Actions for Accidental Releases (1984).

[134] CRAMRA 1988, Article 8.

[135] ILC Draft Liability, Article 24, 26, 28.

[136] Space Liability Convention 1972, Article II.

[137] Chorzow Factory Case (1927) PCIJ ser. A, No. 17, at 47.

[138] Rainbow Warrior Case, 82 I.L.R (1990) 500, 575–577.

[139] Corfu Channel Case ICJ Rep. 1949, 4, 35.

[140] UNCLOS 1982, Article 139, 235.

[141] See: (Barboza 1990), p. 39.

[142] Stockholm Declaration 1972, Principle 22: Rio Declaration 1992, Principle 13.

The Caspian Sea Convention of 2018 envisages the liability of the polluting party for damage caused to the environmental ecological system of the Caspian Sea (Article 3, Para. 13).

10.8.3 Settlement of Disputes

There is a remarkably constant increase in the frequency of dispute settlement clauses among the international legal instruments. The dispute settlement provisions are often contained in multilateral environmental treaties; however, they can also derive from other general agreements regarding the peaceful settlement of international disputes when the states involved in a dispute concerning the implementation or interpretation of environmental treaties among them are also parties to that general agreement.[143] Most of the multilateral treaties regarding environmental issues, which include dispute settlement clauses, reflect basic methods for a peaceful settlement of disputes deriving from Article 33(1) of the United Nations Charter. The most comprehensive and complex dispute settlement provision is included in UNCLOS; however, there is no certainty that UNCLOS is applicable to the Caspian Sea.

Based on international legal rules, the Tehran Convention requires the littoral states to settle disputes regarding its application or interpretation through consultation, negotiation, or any other peaceful means of their choice (Article 30). As observers had anticipated, the Tehran Convention has been prepared and concluded under the auspices of UNEP, and its dispute settlement clauses are almost a literal replication of the provisions regarding dispute settlement contained in UNEP regional treaties.[144] States are required to take advantage of, first, diplomatic means and, second, legal means of settlement, though recently there has been a remarkable increase of noncontentious mechanisms. By resorting to diplomatic means, parties to a dispute finally decide to accept or reject a proposed solution. The primary and nonbinding instrument is negotiation between involved states, to identify the conflict and reach a common acceptable outcome.

The practice of "good offices" is applied when diplomatic relations between states have been broken, and it is aimed at bringing the parties to negotiations. However, its usefulness in environmental disputes seems to be rather insubstantial. Mediation involves an intervention by a third party to settle the dispute while advancing nonbinding proposals made by this third party. The objective of the next settlement instrument inquiry is to clarify disputed issues found under Article

---

[143]Revised General Act for the Pacific Settlement of International Disputes, adopted by the General Assembly of the United Nations, New York, 2 April 1949, 71 UNTS, 102–127.

[144]Barcelona Convention, Article 28.1; Abidjan Convention 1981, Article 24.1; Cartagena Convention 1983, Article 23.1; Nairobi Convention 1985, Article 24.1; Noumea Convention 1986, Article 26.1.

9 of The Hague Convention on the Pacific Settlement of International Disputes. The last of the diplomatic means of settlement is conciliation, which includes attributes of both inquiry and mediation.[145]

The legal means of dispute settlement result in legally binding decisions for the parties to the dispute and include arbitration and judicial settlement. International arbitration is aimed at the settlement of state disputes by judges chosen by them.[146] International environmental disputes may also be taken to an international court competent to give a legally binding decision.[147] to in force yet.

According to the additional protocols to the Tehran Convention, these disputes between the contracting parties concerning the application and interpretation of the provisions of these protocols shall be settled in accordance with the measures for the settlement of disputes foreseen in the Tehran Convention.

10.9 Conclusion

The rules for the protection of the marine environment, which include the prevention, reduction, and control of pollution, as well as the protection, preservation, and restoration of the marine environment, belong to the most highly developed principles in the field of environmental law. As this paper demonstrates, the Tehran Convention contains a set of regulatory methods approaching pollution from different sources, including land-based sources, seabed activities, vessels, dumping, other human activities, introduction of invasive alien species, as well as environmental emergencies. Addressing a number of pollution sources, the Tehran Convention refers to many internationally recognized rules, often of considerable specificity, pertaining to different bodies of water. An important goal of this paper has been also to provide evidence that the Tehran Convention meets all of the main internationally recognized standards relating to environmental protection. This paper has questioned and demonstrated, point by point, that the legal and institutional structures of the Tehran Convention for the Protection of the Marine Environment of the Caspian Sea meet the objective of protecting the fragile marine environment of the Caspian Sea.

---

[145]Revised General Act for the Pacific Settlement of International Disputes (1949), Article 15.1.

[146]Pacific Settlement of International Disputes (1907), Article 37.

[147]International Court of Justice; Chamber for Environmental Matters, European Court of Justice; Court of First Instance; Human Rights Courts: European Court of Human Rights, Inter-American Court of Human Rights; International Tribunal for the Law of the Sea; Special Arbitral Tribunal; United Nations Conference on Environment and Development; non-governmental "international courts."

References

Barannik V, Borysova O, Stolberg F (2004) The Caspian Sea region: environmental change. Ambio

Barboza J (1990) Sixth report. UN Doc, Geneva

Coordination Committee for Hydrometeorology of the Caspian Sea (2005) 17th COORDINATION COMMITTEE SESSION HYDROMETEOROLOGY AND POLLUTION MONITORING CASPIAN SEA. [Online] Available at: http://www.caspcom.com/index.php?razdsess&lang1&sess17&podsess52. Accessed 18 July 2020

Davidson J (2003) Tomorrow's standing today: how the equitable jurisdiction clause of Article III, Section 2 confers standing upon future generations. Columbia J Environ Law 28:185

Farber D (2003) From here to eternity: environmental law and future generations. Univ Ill Law Rev 289

Fleury R (1995) Das Vorsorgeprinzip im Umweltrecht. Heymann, Köln, Berlin

Golubev G (1997) Environmental problems of large central Asian lakes. In: Jorgenson EME (ed) Worlds' lakes in crises. Brassey's, London, Washington

Hekimoğlu I (1999) Caspian oil and the environment. In: Croissant M, Aras B (eds) Oil and geopolitics in the Caspian Sea Region. Praeger, Westport

Holstein A, Kappas M, Propastin P, Renchin T (2018) Oil spill detection in the Kazakhstan sector of the Caspian Sea with the help of ENVISAT ASAR data. Environ Earth Sci 77

Land K (2000) Souveränität und friedliche Streitbeilegung. Europäische Hochschulschriften. 2950s ред. Peter Lang (б.м.)

McAllister S (1998) Human rights and the environment. Colorado J Int Law Yearb 9:187

O'connell J, Oldfather CM (1993) A lost opportunity: a review of the American Law Institute's reporters' study on enterprise responsibility for personal injury. An Diego Law Rev Assoc 30:307

Palast G (2012) BP Covered Up Blow-out Two Years Prior to Deadly Deepwater Horizon Spill. [Online] Available at: https://www.ecowatch.com/bp-covered-up-blow-out-two-years-prior-to-deadly-deepwater-horizon-spi-1881610168.html. Accessed 18 July 2020

Thompson B (2004) The trouble with time: influencing the conservation choices of future generations. Nat Res J 44:601

Tilleman W (1995) Public participation in the environmental impact assessment process: a comparative study of impact assessment in Canada, the United States and the European Community. Columbia J Transnatl Law 33:337–439

Chapter 11
Concluding Remarks

The political opening of the Caspian region after the collapse of the Soviet Union brought the use of the rich oil and gas resources of the Caspian Sea to the fore of regional and global politics. As a result of the dissolution of the USSR, five instead of two coastal countries emerged in the Caspian region. Moreover, their state borders were not settled. The revenue prospects from the use of the natural resources of the Caspian Sea supported the economic recovery of the riparian states. On the other hand, it hardened the states' negotiating positions and their lack of willingness to compromise regarding demarcating borders, as well as in defining the legal regime in the Caspian Sea. The potential for conflict that arises from the growing demand of coastal states for fossil fuels, sharpened by the uncertainty of the legal situation in the Caspian region, met with the divergent interests of global players such as the United States, the EU, and China, which tried to gain access to Caspian resources. There is plain danger in the fact that the slow progress in multilateral negotiations on access to transboundary energy fields could replace the search for a legal compromise with the expanded armament of the Caspian's littoral states. This in turn could lead to a paralysis of regional cooperation and economic development—as well as to a weakening of regional security.

The legal investigation carried out as part of this research refers on the one hand to a number of unresolved legal issues in the Caspian Sea, ones that were mainly caused by the coastal states' lack of political will to compromise. At the same time, it presents the progress in clarifying the legal relationships in the region and thus the raising of hope for the peaceful and mutually satisfying settlement of the legal conflict in the region. It is worth stating that since the late 1990s, certain steps have been achieved in the multilateral negotiations on the draft multilateral agreement on the legal status of the Caspian Sea. In addition, some Caspian Sea countries have completed bilateral contracts clarifying the regime for the use of the energy sources in other parts of the Caspian Sea. New legal frameworks are still coming into being, and thus to achieve success, the negotiating parties need to increase their flexibility, be willing to compromise, and abandon preconditions.

In the perspective of global energy security, the natural resources of the Caspian Sea play a rather minor role, but for the newly independent states of Azerbaijan, Kazakhstan, and Turkmenistan, they are the only reliable guarantee of economic recovery and development. At the same time, the availability of natural resources ensures a basis for the independent political development of these countries and their regional position. Thus, the Caspian resources have an influence not only on the regional economy as they also have a strong impact on the political stability in the region. Conversely, a political crisis in the Caspian region—ethnic, religious, ecological, or territorial—may cause a high degree of regional destabilization as well as affect regional energy security. The question of security requires solving the problem of protection of the regional environment and respect for human rights.

All these require the establishment of a new supranational legal framework in the region. The existing regional conflicts result mainly from the undefined legal status of the Caspian Sea and from the lack of unity among the coastal countries concerning the allocation of Caspian resources. Since the collapse of the Soviet Union, the problem of the international legal status and regime of the Caspian Sea has remained unresolved. Despite ongoing negotiations between the riparian countries, a certain status quo was established that does not allow accelerating settlement. The Soviet–Iranian Treaties of 1921 and 1940 remaining in force are incomplete because except for fishing and navigation, they do not regulate any other regime for the use of the Caspian Sea. The interpretation of the existing status (whether the Caspian Sea is a sea or a lake or a condominium in the legal sense) does not allow drawing respective legal conclusions regarding the use of Caspian resources. Instead of continuing this long-standing, fruitless discussion of whether the Caspian is a sea or a lake in a legal sense, the emphasis of the current discourse over the status of the Caspian Sea is restricted to three groups of normative acts: the first, most recent group is that of bilateral treaties concluded between Russia, Kazakhstan, and Azerbaijan, delimiting sectors of the seabed and subsoil in the northern part of the Caspian Sea for exploiting natural resources. In spite of the denial of their legal power by Iran and Turkmenistan, the treaties contribute to the clarification of the current regime of the use of Caspian natural resources without prejudicing the future shape of the Caspian Sea's legal status.

The second notable milestone in the current international legal development in the Caspian region is the signing in 2003 of the Tehran Convention on environmental protection in the Caspian Sea. The adoption of this treaty, being the only document signed after the collapse of the Soviet Union by all littoral states, reflects the states' awareness of the need to strengthen the protection of the fragile Caspian environment. Its full operation will be achieved, however, with the signature of two of the four remaining additional protocols.

The conclusion of the Tehran Convention as well as the signing of the North Caspian Agreements between Russia, Kazakhstan, and Azerbaijan reflect new intrastate legal developments. Despite the statement that these agreements do not undermine the importance of multilateral negotiations on the legal status of the Caspian Sea, their practical impact upon negotiations is plain. All these recent agreements, which in their content focus on separate legal regimes in using and protecting of the Caspian Sea, they represent a new approach to solving legal regional uncertainties by gradually solving separate issues instead of awaiting the final settlement of the entire status of the Caspian Sea. It is still argued, but not any more practically followed, that any regulation of Caspian affairs must be taken with the consent of all coastal states. Obviously, the final status of the Caspian Sea requires its settlement by a unanimous decision of all the coastal states in the form of a Convention on the legal status of the Caspian Sea, including a comprehensive regulation of all legal issues. In the meantime, it seems to have become acceptable that decisions on certain legal regimes are taken separately and incrementally.

The multilateral negotiations of the riparian states over the future Convention on the legal status of the Caspian Sea are the last but not the least of the three aspects of the current discourse over the future status of the Caspian Sea. Founded in 1995, the working group of deputy foreign ministers of all five littoral states provided a mechanism for continuous negotiations concerning the legal status of the Caspian Sea. In the process of its work, the group developed a Draft Caspian Status Convention. The draft refers to all aspects of the legal regime and the status of the Caspian Sea—including demarcating borders, the regulation of fishing, navigation, the use of natural resources, etc.—and thus in comparison to sectoral agreements, this represents an exhaustive legal regulation. Due to its preliminary status, the Draft Convention has no binding force upon the coastal states, but its provisions reflect a clearly positive legal development in the Caspian region mainly governed by international legal standards. Upon obtaining the consent of all coastal states, the Draft Convention will be binding upon the coastal states and will become part of the public law of the sea. Its provisions, which nowadays represent merely a certain tendency in the legal development of the Caspian region in the future, will turn into binding rules, and thus following its development is of practical importance today as well to enable representatives of the law practice to prepare for the upcoming changes.

The existing clash of energy interests among the Caspian coastal states, sharpened by the permanent interference of international actors, hampers the prognosis of future developments in the geopolitical situation in the Caspian region. The situation is further complicated by the prevailing lack of clarity concerning the legal situation in the region, where more than 20 years after the collapse of the Soviet Union the international legal status of the Caspian Sea is still undefined. The uncertainty of the maritime delimitation between the riparian states and, thus, the undefined scope of their sovereign rights on natural resource exploitation destabilize the economic and political situation in the Caspian region.

The ongoing legal debate on the existing status of the Caspian Sea, both by regional lawyers and lawyers in Western countries, is limited to the analysis of relevant agreements between the Soviet Union and Persia/Iran and to the related question as to whether the Caspian is a sea or a lake. However, there is a need to extend this discussion to include the latest legal developments—such as the conclusion of the Northern Caspian Agreements and the Tehran Convention—and to try to analyze their impact on the existing legal framework in the region. The investigation of the Draft Convention on the future status of the Caspian Sea allows insight into the possible future legal developments—and thus more comprehensive legal analysis of the existing situation. Both analyses are *sine qua non* if the legal security in the Caspian Sea is to increase and if closer cooperation of its littoral states is to be secured.

List of International Treaties

Aarhus Convention (Convention on Access to Information, Public Participation in Decision-Making and Access to Justice in Environmental Matters of 1998)
Abidjan Emergency Protocol of 1981 (Abidjan Protocol Concerning Co-operation in Combating Pollution in Cases of Emergency of 1981)
African Convention on the Conservation of Nature and Natural Resources of 1968
African Phyto-Sanitary Convention of 1954 (Phyto-sanitary Convention for Africa South of the Sahara of 1954)
Agreement establishing the EBRD of 1990 (Agreement Establishing the European Bank for Reconstruction and Development of 1990)
Agreement for the Establishment of a General Fisheries Council for the Mediterranean of 1949
Agreement on Implementation of the Provisions of the United Nations Convention of 10.12.1982 on the Conservation and Management of Straddling Fish Stocks and Highly migratory Fish Stocks
Agreement concerning Co-operation in Marine Fishing of 1962
Alpine Convention of 1991
Amazonian Treaty 1978 (Amazon Cooperation Treaty of 1978)
Anglo-French Conventions of 1898 & 1904 & 1906 concerning the Lake Chad (Chad, Cameroon, Niger, Nigeria)
Anglo-German Agreement of 1890
Anglo-German Agreement of 1890 (concerning the Lake Victoria, Uganda, Kenya, Tanzania)
Anglo-Portuguese Agreement of 1954 concerning the Lake Malawi
Antarctic Marine living Resources Convention 1980 (Convention on the Conservation of Antarctic Marine Living Resources of 1980)
Antarctic Seals Convention 1972 (Convention for the Conservation of Antarctic Seals (CCAS) of 1972)
Antarctic Treaty of 1959
Antarctic-Environmental Protocol (Protocol on Environmental Protection to the Antarctic Treaty of 1991)

Asbestos Convention of 1986
ASEAN Convention of 1985 (ASEAN Agreement on the Conservation of Nature and Natural Resources of 1985)
Athens LBS Protocol (Protocol for the Protection of the Mediterranean Sea against Pollution from Land-Based Sources (LBS Protocol) of 1980)
Baltic Fishing Convention 1973 (Convention on Fishing and Conservation of the Living Resources in the Baltic Sea and the Belts of 1973)
Bamako Convention of 1991 (Bamako Convention on the ban on the Import into Africa and the Control of Transboundary Movement and Management of Hazardous Wastes within Africa of 1991)
Barcelona Convention of 1976 (Barcelona Convention for Protection against Pollution in the Mediterranean Sea of 1976)
Barcelona Dumping Protocol (Protocol for the Prevention of Pollution of the Mediterranean Sea by Dumping from Ships and Aircraft of 1976)
Basel Convention of 1989 (Basel Convention on the Control of Transboundary Movements of Hazardous Wastes and their Disposal of 1989)
Biodiversity Convention (Convention on Biological Diversity of 1992)
Black Sea Fishing Convention 1959 (Convention concerning Fishing in the Black Sea of 1959)
Bonn Agreement of 1983 (Agreement for cooperation in dealing with pollution of the North Sea by oil and other harmful substances of 1983)
Bonn Convention (Convention on the Conservation of Migratory Species of Wild Animals of 1979)
Border Treaty between Yugoslavia and Greece of 21 May 1959 (concerning Lake Doyran)
Cartagena Convention (Protocol to the Convention for the Protection and Development of the Marine Environment of the Wider Caribbean Region Concerning Co-operation in Combating Oil-Spills in the Wider Caribbean Region of 1983 also known as Oil Spill Protocol)
Cartagena Convention of 1983 (Cartagena Convention for the Protection and Development of the Marine Environment in the Wider Caribbean Region)
CITES (Convention on International Trade in Endangered Species of Wild Fauna and Flora, also known as the Washington Convention of 1973)
COLREG (Convention on the International Regulations for Preventing Collisions at Sea of 1972)
Convention between Persia and Russia of 1881: Convention between Persia and Russia of 9 December 1881
Convention between Russia and Persia of 27 May 1893 for the Territorial Interchange of Faruze in Khorassan
Convention between Switzerland and France on the Determination of the frontier in Lake Geneva of 25th February 1953
Convention for the Prevention of Marine Pollution from Land-Based Sources (Paris Convention on land-based sources of marine pollution of 1974).
Convention for the Protection of Submarine Cables of 1884
Convention of Peking of 1860 (concerning the Lake Khanka)

List of International Treaties 209

Convention on Conservation of North Pacific Fur Seals of 1976
Convention on Fishing and Conservation of Living Resources of the High Seas of 1958 (Geneva)
Convention on Freedom of Transit of 1921
Convention on the Conservation of European Wildlife and Natural Habitats of 1979
Convention on the Continental Shelf of 1958 (Geneva)
Convention on the High Seas of 1958 (Geneva)
Convention on the Territorial Sea and the Contiguous Zone of 1958 (Geneva)
Convention on Transit of Land-Locked States, 1965
Convention on Conduct of Fishing Operations in the North Atlantic of 1967
Copenhagen Agreement of 1971
CRAMRA 1988 (Convention on the Regulation of Antarctic Mineral Resource Activities of 1988)
Cromer–Ghali Agreement of 1899 (Condominium of Sudan)
Danube Fishing Convention 1958 (Convention concerning Fishing in the Waters of Danube of 1958)
English–Belgian Protocol of 1924 (English–Belgian Protocol of 5th August 1924 concerning Lake Tanganyika) (Tanzania, Burundi, Congo)
ENMOD Convention (Convention on the Prohibition of Military or Any Other Hostile Use of Environmental Modification Techniques of 1977, also known as Environmental Modification Convention)
Espoo Convention (Espoo Convention on Environmental Impact Assessment in a Transboundary Context of 1991)
EURATOM Treaty of 1982 (Treaty establishing the European Atomic Energy Community of 1982)
European Plant Protection Convention of 1951 (Convention for the Establishment of the European and Mediterranean Plant Protection Organization of 1951)
Florence Protocol of 1926 (concerning the Lake Ohrid between Yugoslavia and Albania)
Florence Protocol of 1926 (concerning the Lake Prespa between Yugoslavia and Greece)
Florence Protocol of 1926 (concerning the Lake Skadar between Yugoslavia and Albania)
Gastein Convention of 1865
GATT Convention (General Agreement on Tariffs and Trade) of 1994
Global Programme of Action for Protecting the Marine Environment from Land-Based Activities of 1995
Haines-Fairbanks Oil Pipeline Agreement of 1955
Helsinki Convention (Convention on the Protection and Use of Transboundary Watercourses and International Lakes of 1992)
Helsinki Convention of 1992 (Convention on the Protection of the Marine Environment of the Baltic Sea Area of 1992)
IAEA Notification Convention of 1986 (Convention on Early Notification of a Nuclear Accident of 1986)
ILO Radiation Convention of 1960 (ILO Radiation Protection Convention of 1960)

Industrial Accidents Convention of 1992 (Convention on the Transboundary Effects of Industrial Accidents of 1992)
International Convention for the prevention of pollution of the sea by oil of 1954
International Whaling Convention of 1946 (The International Convention for the Regulation of Whaling of 1946)
Jeddah Convention 1982 (Regional Convention for the Conservation of the Red Sea and Gulf of Aden Environment of 1982)
Jeddah Pollution Emergency Protocol of 1982 (Jeddah Protocol Concerning Regional Co-Operation in Combating Pollution by Oil and Other Harmful Substances in Cases of Emergency of 1982)
Kuwait Convention of 1978 (Kuwait Regional Convention for Co-operation on the Protection of the Marine Environment from Pollution of 1978)
Kuwait Emergency Protocol of 1978 (Kuwait Protocol Concerning Co-operation in Combating Pollution by Oil and Other Harmful Substances in Cases of Emergency of 1978)
Kuwait LBS Protocol (Protocol for the Protection of the Marine Environment Against Pollution from Land-Based Sources of 1990)
Lapas Protocol (on Lake Titicaca between Peru and Bolivia) from 2nd June 1925 and 13th February 1932
Lima Convention of 1981 (Convention for the Protection of the Marine Environment and Coastal Area of the South-East Pacific of 1981)
Load Line Convention of 1966
London Agreement of 1915 (concerning Lake Albert) of 3rd February 1915 between Belgium and Great Britain
London Convention of 1933 (Convention Relative to the Preservation of Fauna and Flora in the Natural State)
London Dumping Convention (Convention on the Prevention of Marine Pollution by Dumping of Wastes and Other Matter of 1972)
Lugano Convention (Convention on Civil Liability for Damage Resulting from Activities Dangerous to the Environment of 1993)
Luso-British Agreement of 1891
MARPOL (International Convention for the Prevention of Pollution From Ships, 1973 as modified by the Protocol of 1978)
Mediterranean Emergency Protocol of 1976 (Protocol Concerning Cooperation in Combating Pollution of the Mediterranean Sea by Oil and other Harmful Substances in Cases of Emergency of 1976)
Montreal Protocol of 1987 (Montreal Protocol on Substances that Deplete the Ozone Layer of 1987)
Moscow Peace Treaty of 1940 (concerning the Lake Ladoga)
NAFTA Agreement (North American Free Trade Agreement) of 1994
Nairobi Convention of 1985 (Nairobi Convention of the Protection, Management and Development of the Marine and Coastal Environment of the Eastern African Region of 1985)
New York Convention of 1965 (New York Convention on Transit Trade of Land-locked States of 1965)

List of International Treaties 211

North Atlantic Salmon Convention 1982 (Convention for the Conservation of Salmon in the North Atlantic Ocean of 1982)
North-East Atlantic Fisheries Convention of 1959
Northern Gas Pipeline Agreement, 1977
Northwest Atlantic Fisheries Convention 1978 (Convention on Future Multilateral Cooperation in the Northwest Atlantic Fisheries of 1978)
Noumea Convention of 1986 (Noumea Convention for the Protection of Natural Resources and Environment of the South Pacific Region of 1986)
Noumea Dumping Protocol (Protocol for the Prevention of Pollution of the South Pacific Region by Dumping of 1986)
Occupational Health Services Convention of 1985
OPRC (Oil Pollution Preparedness Convention) (International Convention on Oil Pollution Preparedness, Response and Co-operation of 1990)
Oslo Convention (Convention for the Prevention of Marine Pollution by Dumping from Ships and Aircraft of 1972)
OSPAR Convention (Convention for the Protection of the Marine Environment of the North-East Atlantic of 1992)
Pacific Settlement of International Disputes 1907 (Hague Convention for the Pacific Settlement of International Disputes of 1907)
Paipa Dumping Protocol (Protocol for the Conservation and Management of Protected Marine and Coastal Areas of the South-East Pacific of 1989)
Plant Protection Agreement of 1956 (Plant Protection Agreement for the Asia and Pacific Region of 1956)
Intervention Protocol of 1973 (Protocol Relating to Intervention on the High Seas in Cases of Pollution by Substances Other Than Oil of 1973)
Quito LBS Protocol (Protocol for the Protection of the South-East Pacific against Pollution from Land-based Sources of 1983)
Ramsar Wetlands Convention 1971 (Convention on Wetlands of International Importance, especially as Waterfowl Habitat of 1971)
Rhine Chemical Pollution Convention (Convention For The Protection Of The Rhine Against Chemical Pollution of 1976)
Rotterdam Convention 1998 (Rotterdam Convention on the Prior Informed Consent Procedure for Certain Hazardous Chemicals and Pesticides in International Trade of 1998 also known as Prior Informed Consent)
Rotterdam Convention of 1998 (also known as Prior Informed Consent).
SAR Convention (International Convention on Maritime Search and Rescue, 1979)
SOLAS (International Convention for the Safety of Life at Sea of 1974)
South Atlantic Fisheries Convention 1969 (Convention on the Conservation of the Living Resources of the Southeast Atlantic of 1969)
South Pacific Nature Convention of 1976 (Convention on Conservation of Nature in the South Pacific of 1976)
South-West Asia Locust Agreement of 1963 (Agreement for the Establishment of a Commission for Controlling the Desert Locust in the Eastern Region of its Distribution Area in South-West Asia of 1963)

Space Liability Convention of 1972 (Convention on International Liability for Damage Caused by Space Objects of 1972)
Switzerland–Italy Agreements concerning the boundary in Lake Lugano
Treaty between Brazil and Uruguay Modifying their Frontiers on Lake Mirim and the River Yaguaron, and Establishing General Principles of Trade and Navigation in those Regions of 1909
Treaty of Rarotonga (South Pacific Nuclear-Free Zone Treaty of 1985)
Treaty of Trianon (Treaty of Peace Between The Allied and Associated Powers and Hungary And Protocol and Declaration of 1920), concerning the boundary of Neusiedler See
Tropical Tuna Commission Convention of 1949 (Convention for the Establishment of an Inter-American Tropical Tuna Commission of 1949)
UNCLOS (United Nations Convention on the Law of the Sea of 1982)
UNFCCC (United Nations Framework Convention on Climate Change of 1992)
United Nations Charter of 1945
United Nations Convention on a Code of Conduct for Liner Conferences of 1974
United Nations Convention on Conditions for Registration of Ships of 1986
Vienna Convention of 1985 (Vienna Convention for the Protection of the Ozone Layer of 1985)
Vienna Convention on Succession of States in respect of Treaties, 1978
Vienna Convention on the Law of Treaties 1969
UN Water Convention (Convention on the Law of the Non-navigational Uses of International Watercourses of 1997.
Western Hemisphere Convention of 1940 (Convention on Nature Protection and Wild Life Preservation in the Western Hemisphere of 1940)

Bibliography

Carleton C (1994) The Maritime Boundary Agreements between the United Kingdom and United States of America. IJMCL 9(2):258–259

Grabitz E, Hilf M (2008) Kommentar zur Europäischen Union, 2nd edn. CH Beck, München

Kembayev Z (2008) Die Rechtslage des Kaspischen Meeres. Zeitschrift für ausländisches öffentliches Recht und Völkerrecht, 68

MacEven A (1971) International boundaries of East Africa. Cambridge University Press, Oxford

Müller F, Halbach U (2001) Persischer Golf, Kaspisches Meer und Kaukasus – Entsteht eine Region vitalen europäischen Interesses. SWP-Studie, Berlin

Okidi C (1980) Legal and policy regime of Lake Victoria and Nile Basins. Int J Law Libr 20 (3):395–447

Peric M (1977) The problem of boundaries in Africa. ред. Belgrade: Rev Int Aff 28

Shapiro L (1950) Soviet Treaty series: a collection of bilateral treaties, agreements and Conventions, etc. concluded between the Soviet Union and Foreign powers. Georgetown University Press, Washington, DC

Tracey G (2014) Russia and the Caspian Sea: projecting power or competing for influence. United States Army War College Press, Carlisle Barras

Yury M (1999) Legal status of the Caspian Sea. Int Aff Russian J World Polit Diplomacy Int Relat 45(1)

Lightning Source UK Ltd.
Milton Keynes UK
UKHW020749280722
406505UK00001B/2